Primarily Earth

Principle Authors

Evalyn Hoover   Sheryl Mercier

Contributing Authors

Carol Gossett   Norma Hashimoto
Ann Harris   Jody Irola

Meta Thomas

Editors

Betty Cordel   Judith Hillen

Illustrator

Sheryl Mercier

Desktop Publisher

Roxanne Williams

Primarily Earth has been developed through a cooperative learning program involving the AIMS Education Foundation and Fresno Unified School District.

This book contains materials developed by the AIMS Education Foundation. **AIMS** (**A**ctivities **I**ntegrating **M**athematics and **S**cience) began in 1981 with a grant from the National Science Foundation. The non-profit AIMS Education Foundation publishes hands-on instructional materials (books and the monthly *AIMS* magazine) that integrate curricular disciplines such as mathematics, science, language arts, and social studies. The Foundation sponsors a national program of professional development through which educators may gain both an understanding of the AIMS philosophy and expertise in teaching by integrated, hands-on methods.

ISBN 1-881431-63-0

Printed in the United States of America

I HEAR AND I FORGET

I SEE AND I REMEMBER

I DO AND I UNDERSTAND

—Chinese Proverb

Table of Contents

INTEGRATED PROCESSES

	Observing	Communicating	Comparing /contrasting	Collecting/recording data	Classifying	Predicting	Generalizing	Applying	Inferring	Measuring	Graphing
The Earth's Features	X	X	X	X							
My Rock	X	X	X	X	X	X				X	
Rock Groups	X		X	X	X		X			X	X
Rock and More Rocks	X		X	X	X		X			X	X
Ice Breakers	X	X		X		X	X	X			
Agent Erosion	X		X	X		X		X			
Soil Study	X	X	X	X					X		X
Sandpile	X	X	X	X	X						
The Earth Has What We Need!	X	X	X	X	X				X		
What's Inside?	X		X	X	X				X		
Quaking Earth	X	X		X	X	X		X			
Volcanoes	X	X		X				X			
Where is Water?	X	X	X						X		
What Makes Rain?	X	X		X	X	X		X			
A Disppearing Act	X		X	X		X	X			X	
Water to Ice to Water	X		X	X							
A Close Look at Air	X	X	X				X	X			
Air is There	X	X	X			X	X	X			
The Wind Blows	X		X	X	X					X	
Which Way?	X				X	X		X			
Cloudy Weather	X	X	X		X	X			X		
Watching the Weather	X		X	X		X			X		X
Air Temperature	X	X	X	X		X				X	

v

Project 2061 Benchmarks*

AIMS is committed to remaining at the cutting edge of providing integrated math/science studies that are user friendly, educationally sound, developmentally appropriate, and aligned with the recommendations from national education documents.

Below you will find a listing of the *Benchmarks for Science Literacy* (American Association for the Advancement of Science) which are addressed in this publication.

- *Everybody can do science and invent things and ideas.*
- *A model of something is different from the real thing but can be used to learn something about the real thing.*
- *People can often learn about things around them by just observing those things carefully, but sometimes they can learn more by doing something to the things and noting what happens.*
- *Change is something that happens to many things.*

- *Tools such as thermometers, magnifiers, rulers, or balances often give more information about things than can be obtained by just observing things without their help.*
- *Objects can be described in terms of the materials they are made of (clay, cloth, paper, etc.) and their physical properties (color, size, shape, weight, texture, flexibility, etc.).*
- *Many materials can be recycled and used again, sometimes in different forms.*
- *Magnets can be used to make some things move without being touched.*
- *People can use objects and ways of doing things to solve problems.*
- *Chunks of rocks come in many sizes and shapes, from boulders to grains of sand and even smaller.*

- *Describing things as accurately as possible is important in science because it enables people to compare their observations with those of others.*
- *Numbers can be used to count things, place them in order, or name them.*
- *Simple graphs can help to tell about observations.*
- *Shapes such as circles, squares, and triangles can be used to describe many things that can be seen.*
- *Often a person can find out about a group of things by studying just a few of them.*
- *One way to describe something is to say how it is like something else.*
- *Things change in some ways and stay the same in some ways.*
- *Often you can find out about something big by studying just a small part of it.*
- *Magnifiers help people see things they could not see without them.*
- *Things in nature and things people make have very different sizes, weights, ages and speeds.*
- *Some events in nature have a repeating pattern. The weather changes some from day to day, but things such as temperature and rain (or snow) tend to be high, low, or medium in the same months every year.*

- We can learn about things around us by just watching carefully or by doing something to them and seeing what happens.
- Water can be a liquid or a solid and can be made to go back and forth from one form to the other. If water is turned into ice and then the ice is allowed to melt, the amount of water is the same as it was before freezing.
- Water left in an open container disappears, but water in a closed container does not disappear.
- People need water, food, air, waste removal and a particular range of temperatures in their environment, just as other animals do.
- The sun warms the land, air and water.
- Most living things need water, food, and air.
- Things can change in different ways, such as in size, weight, color, and movement. Some small changes can be detected by taking measurements.
- Some changes are so slow or so fast that they are hard to see.
- Raise questions about the world around them and be willing to seek answers to some of them by making careful observations and trying things out.

- Tools are used to do things better or more easily and to do some things that could not otherwise be done at all. In technology, tools are used to observe, measure, and make things.
- Some materials can be used over again.
- Waves, wind, water, and ice shape and reshape the earth's land surface by eroding rock and soil in some areas and depositing them in other areas, sometimes in seasonal layers.

- In doing science, it is often helpful to work with a team and to share findings with others. All team members should reach their own individual conclusions however, about what the findings mean.
- Things move in many different ways, such as straight, zigzag, round and round, back and forth, and fast and slow.
- Some things are more likely to happen than others. Some events can be predicted well and some cannot. Sometimes people aren't sure what will happen because they don't know everything that might be having an effect.

vii

"From their very first day in school, students should be actively engaged in learning to view the world scientifically. That means encouraging them to ask questions about nature and to seek answers, collect things, count and measure things, make qualitative observations, organize collections and observations, discuss findings, etc. Getting into the spirit of science and liking science are what count most. Awareness of the scientific world view can come later."(p.6)

"There are many ways to acquaint children with earth-related phenomena that they will only come to understand later as being cyclic. For instance, students can start to keep daily records of temperature (hot, cold, pleasant) and precipitation (none, some, lots), and plot them by week, month, and years. It is enough for students to spot the patterns of ups and downs, without getting deeply into the nature of climate. They should become familiar with the freezing of water and melting of ice (with no change in weight), the disappearance of wetness into the air, and the appearance of water on cold surfaces. Evaporation and condensation will mean nothing different from disappearance and appearance, perhaps for several years, until students begin to understand that the evaporated water is still present in the form of invisibly small molecules."(p.67)

"Teaching geological facts about how the face of the earth changes serves little purpose in these early years. Students should start becoming familiar with all aspects of their immediate surroundings, including what things change and what seems to cause change." (p.72)

* American Association for the Advancement of Science. **Benchmarks for Science Literacy**. Oxford University Press. NY. 1993.

National Science Education Standards*

Below you will find a listing of the National Science Education Standards (National Research Council) which are addressed in this publication.

- *Ask a question about objects, organisms, and events in the environment.*
- *Employ simple equipment and tools to gather data and extend the senses.*
- *Communicate investigations and explanations.*
- *Simple instruments, such as magnifiers, thermometers, and rulers, provide more information than scientists obtain using only their senses.*
- *Objects have many observable properties, including size, weight, shape, color, temperature, and the ability to react with other substances. Those properties can be measured using tools, such as rulers, balances, and thermometers.*
- *Materials can exist in different states—solid, liquid, and gas. Some common materials, such as water, can be changed from one state to another by heating or cooling.*
- *Earth materials are solid rocks and soils, water, and the gases of the atmosphere. The varied materials have different physical and chemical properties, which make them useful in different ways, for example, as building materials, as sources of fuel, or for growing the plants we use as food. Earth materials provide many of the resources that humans use.*
- *Soils have properties of color and texture, capacity to retain water, and ability to support the growth of many kinds of plants, including those in our food supply.*

- *The sun, moon, stars, clouds, birds, and airplanes all have properties, locations, and movements that can be observed and described.*
- *The surface of the earth changes. Some changes are due to slow processes, such as erosion and weathering, and some changes are due to rapid processes, such as landslides, volcanic eruptions, and earthquakes.*
- *Weather changes from day to day and over the seasons. Weather can be described by measurable quantities, such as temperature, wind direction and speed, and precipitation.*
- *Resources are things that we get from the living and nonliving environment to meet the needs and wants of a population.*
- *Some resources are basic materials, such as air, water, and soil; some are produced from basic resources, such as food, fuel, and building materials; and some resources are nonmaterial, such as quiet places, beauty, security, and safety.*
- *The supply of many resources is limited. If used, resources can be extended through recycling and decreased use.*

* *National Research Council.* **National Science Education Standards.** *National Academy Press. Washington, DC. 1996.*

Primarily Earth

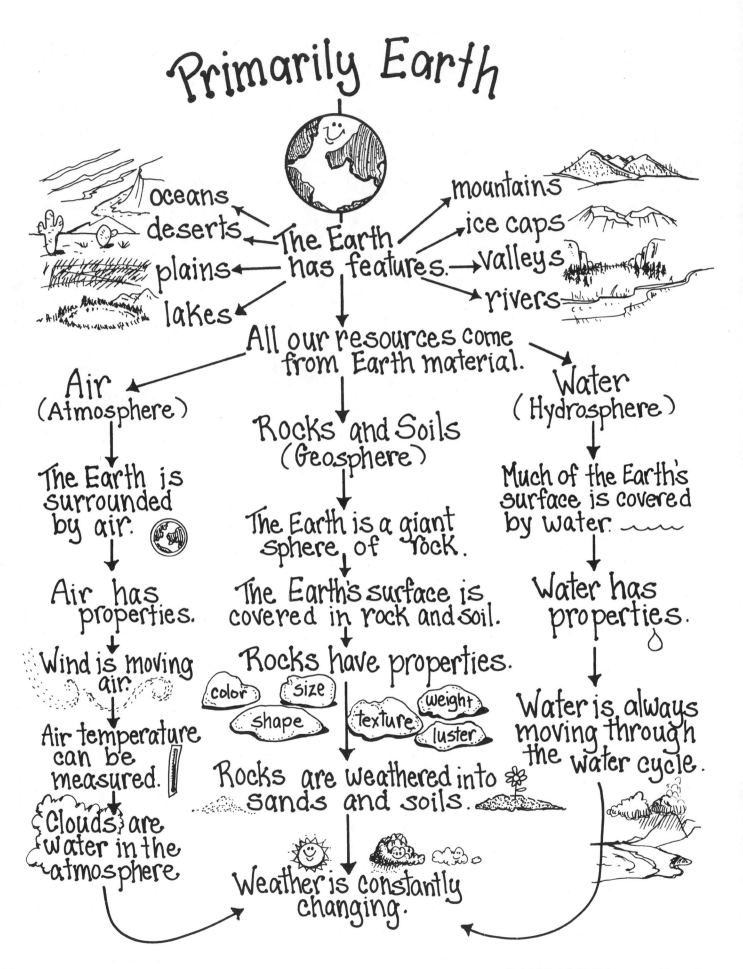

The Earth has features.

oceans
deserts
plains
lakes

mountains
ice caps
valleys
rivers

All our resources come from Earth material.

Air (Atmosphere)

The Earth is surrounded by air.

Air has properties.

Wind is moving air.

Air temperature can be measured.

Clouds are water in the atmosphere

Rocks and Soils (Geosphere)

The Earth is a giant sphere of rock.

The Earth's surface is covered in rock and soil.

Rocks have properties.

color   size
shape   texture   weight   luster

Rocks are weathered into sands and soils.

Weather is constantly changing.

Water (Hydrosphere)

Much of the Earth's surface is covered by water.

Water has properties.

Water is always moving through the water cycle.

Primarily Earth

Overview

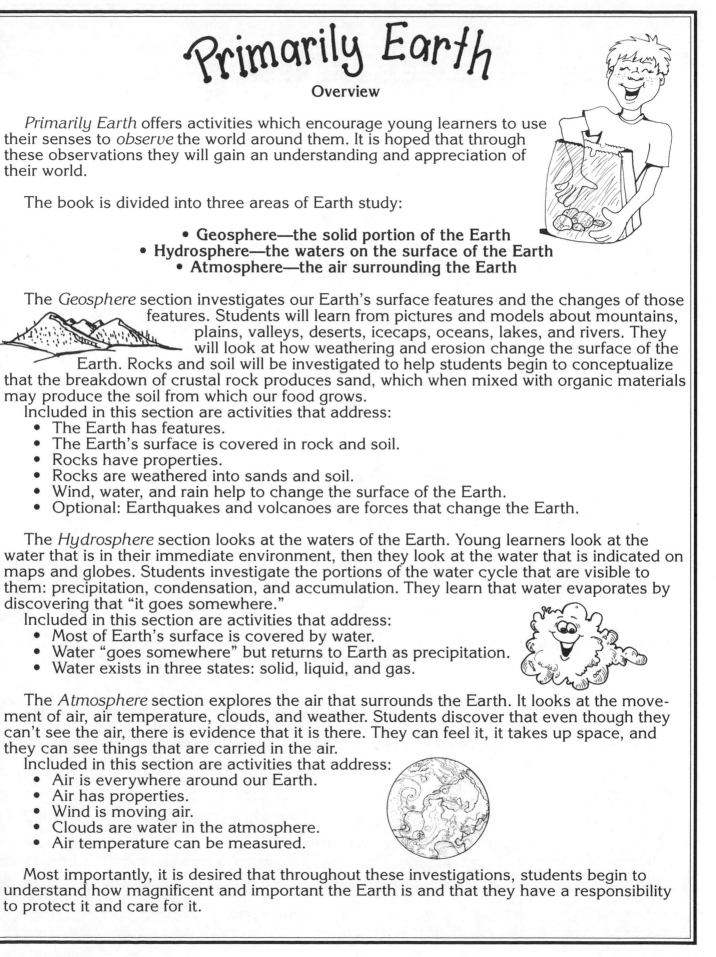

Primarily Earth offers activities which encourage young learners to use their senses to *observe* the world around them. It is hoped that through these observations they will gain an understanding and appreciation of their world.

The book is divided into three areas of Earth study:

- **Geosphere—the solid portion of the Earth**
- **Hydrosphere—the waters on the surface of the Earth**
- **Atmosphere—the air surrounding the Earth**

The *Geosphere* section investigates our Earth's surface features and the changes of those features. Students will learn from pictures and models about mountains, plains, valleys, deserts, icecaps, oceans, lakes, and rivers. They will look at how weathering and erosion change the surface of the Earth. Rocks and soil will be investigated to help students begin to conceptualize that the breakdown of crustal rock produces sand, which when mixed with organic materials may produce the soil from which our food grows.

Included in this section are activities that address:
- The Earth has features.
- The Earth's surface is covered in rock and soil.
- Rocks have properties.
- Rocks are weathered into sands and soil.
- Wind, water, and rain help to change the surface of the Earth.
- Optional: Earthquakes and volcanoes are forces that change the Earth.

The *Hydrosphere* section looks at the waters of the Earth. Young learners look at the water that is in their immediate environment, then they look at the water that is indicated on maps and globes. Students investigate the portions of the water cycle that are visible to them: precipitation, condensation, and accumulation. They learn that water evaporates by discovering that "it goes somewhere."

Included in this section are activities that address:
- Most of Earth's surface is covered by water.
- Water "goes somewhere" but returns to Earth as precipitation.
- Water exists in three states: solid, liquid, and gas.

The *Atmosphere* section explores the air that surrounds the Earth. It looks at the movement of air, air temperature, clouds, and weather. Students discover that even though they can't see the air, there is evidence that it is there. They can feel it, it takes up space, and they can see things that are carried in the air.

Included in this section are activities that address:
- Air is everywhere around our Earth.
- Air has properties.
- Wind is moving air.
- Clouds are water in the atmosphere.
- Air temperature can be measured.

Most importantly, it is desired that throughout these investigations, students begin to understand how magnificent and important the Earth is and that they have a responsibility to protect it and care for it.

Geosphere

Background Information

   Earth is a huge sphere of rock covered by water and soil, and surrounded by air. About 70% of the Earth's surface is covered with water, most of it in the oceans. Land makes up about 30%.

   Earth's crust has a very interesting and varied landscape made up of mountains, valleys, plains, deserts, oceans, lakes, rivers, and icecaps.

- Mountains are large landforms that rise above the surrounding land
- Valleys are depressions on Earth's surface
- Plains are areas of relatively level, treeless land
- Deserts are areas with very little water available
- Oceans are huge bodies of water that cover most of our Earth
- Lakes are bodies of water that form in hollows or basins
- Rivers are streams of water moving across the land
- Icecaps are huge areas of ice found at the North and South Poles

   Geologists investigate materials of Earth's crust such as soil, sand, and rocks. Rocks come in a large variety of shapes, textures, and colors.

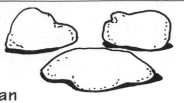

Rocks are solid objects that make up Earth's surface. Rock underlies the hills and mountains; it is under the ocean and icecaps. Layers of rock can be seen where roads cut through hills. Rock outcrops extend above the sand in the desert. Rock can be seen along the shorelines of oceans and lakes. Beneath Earth's surface are the rocks which form the crust. The crust is about 30 miles thick.

Rocks are slowly broken down into smaller pieces by a process called weathering.

One end product of weathering is soil. Weathering is essential in soil formation because soil begins to form when bedrock crumbles into fine particles. Organic material, resulting from the decaying of plant and animal matter (humus), adds richness to the crumbled inorganic particles thus producing soil.

Land is shaped and carved by water, wind, and ice—agents of erosion. Over many years, these erosive forces wear down the soil and rocks and transport them to other places.

Rainwater collects in cracks in rock. When the temperature gets cold enough, the water turns to ice. Water, when it freezes, expands, pushing against the sides of the crack, breaking off small pieces of rock. The rain falling on the land picks up and carries soil and rock fragments.

Wind causes erosion mostly along the ground's surface. It picks up and carries tiny particles of sand. When the sand blows against solid rock, it chips off more bits as it continues to wear away the rock.

Glaciers, moving sheets of ice, cause erosion as they scrape deep hollows out of the land and wear the rock away.

People can cause erosion, too. When crowds of people step on fragile plants, they tend to make paths. These paths of barren soil make it easier for the soil to be blown or washed away.

Landforms Information

Mountains

Mountains are natural landforms that reach high above the surrounding country. Thus, any part of Earth's surface that stands much higher than the surrounding land may be called a mountain. The height of a mountain is usually expressed as the distance that its peak rises above sea level. Mountains are a major feature of the landscape in many areas. The highest mountain in the United States, Alaska's Mount McKinley, is more than 20,000 feet above sea level.

Mountains generally have steep slopes with sharp or slightly rounded peaks. A mountain may be a single peak, such as a lone volcano, or it may be part of a mountain range.

Streams and rivers often carve out valleys in mountains. Some mountain valleys are wide, especially in older mountains; in the younger mountains, they are usually narrower. In time, mountains are worn down to the level of hills and plains.

Mountain ranges are important because they affect the climate and water flow of surrounding regions. Mountains are also important for the plants and animals they support and as a source of lumber and minerals. Much of the world's mineral resources come from mountainous regions.

Mountains are also important recreation areas. Each year, millions of people vacation in mountainous areas to camp, hike, ski, climb, or just to enjoy the fresh air and beautiful views.

Valleys

A valley is a depression in the Earth's surface. Most valleys are formed by the running water of streams and rivers and by the erosion of slopes leading to them. Rock and soil material is eroded from the slopes of mountains and then moved to the valley floor. A valley may also be formed when a long, narrow section of Earth's crust sinks below the surrounding area.

The bottom of the valley is called its floor. Mountain valleys usually have narrow floors, but in low-lying plains, a floor may be several miles wide. The floors of valleys often have very fertile soil which makes excellent farmland. Many different crops are grown on fertile valley floors.

Plains

Plains are areas that are mostly flat. They are large regions that are quite level or have gently rolling hills with grasses and few trees. The interior plains of the United States were formed as mountains and hills but have been gradually worn down by wind, water, and glaciers.

These central plains have relatively good soil and are used for farming. In the United States, wheat, barley, and oats are some of the crops that grow on the plains. Cattle and sheep are raised in these areas, too.

There are low flat plains along the coasts of the United States. These plains were formed by soil deposits. In the past, shallow oceans covered this area. Soil has been deposited by rivers, streams, and the ocean. The soils of the coastal plains are usually very rich. Agriculture is one of the main activities on coastal plains. Cotton, vegetables, and citrus are some of the crops that are grown in these areas.

Deserts

A desert is a region that receives less than 10 inches of precipitation each year. Deserts have many different landscapes and types of soil. Sand covers about one-fourth of the desert land; gravel-covered hills, mountains, and lowlands cover the rest of the desert landscape. Many kinds of plants and animals flourish in the desert. The plants and animals that are able to live in these harsh environments have adapted to these conditions.

Desert plants have adapted to the extreme dryness of the weather in various ways. Some have roots that go down a hundred feet or more, others have roots that spread out horizontally for great distances. Other plants store large amounts of water in their leaves, roots, or stems. Desert plants tend to be sparsely spread out over a large area so they can utilize the small amount of water available. Quite often a desert area will receive

no rainfall for years, then a big storm will release a brief, violent downpour of rain.

A surprisingly large number of animals, insects, reptiles, birds, and spiders live in the desert. Most desert animals are inactive during the day, staying in shady areas or burrows to stay cool. They come out at night to feed and hunt. These animals obtain water from the food they eat and from the few areas that have water holes.

Deserts are some of the hottest places on Earth. The desert land absorbs more heat from the sun than areas which are covered with thick vegetation. Desert temperatures often reach 100° F or higher during the day and may then drop 50° at night.

Many desert soils are rich in minerals and in some areas, oil and natural gas.

Oceans

The ocean is a great body of salty water that covers more than 70 percent of Earth's surface. Almost all the water on Earth is in the ocean. The waters of the ocean form one great connected body; however, the continents divide this great body of water into four major parts. They are, in order of size, the Pacific Ocean, the Atlantic Ocean, the Indian Ocean, and the Arctic Ocean.

The waters of the ocean are moved around by the forces created by Earth's rotation and the heat of the sun. Tides, created by the gravitational pull of the moon (and to a lesser degree, the sun) also affect the movement of the oceans. Water leaves the surface of the ocean by evaporation and is later condensed forming clouds, only to eventually return as precipitation.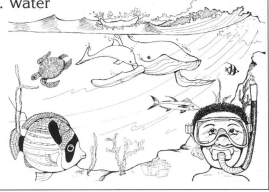

The ocean provides us with many things. It is a source of food, energy, and minerals. The waters of the ocean are used for transporting humans and cargo from port to port. We use the shores of the oceans as a place for swimming, boating, and other forms of recreation.

Lakes

A lake is a body of water surrounded by land. Lakes can be found in all parts of the world. The word *lake* comes from a Greek word meaning hole or pond. Lakes are usually deep depressions in the crust of the Earth that have filled with fresh water. Rain, melting snow, water from springs and rivers, and surface runoff fill these depressions.

The largest number of lakes lies in regions that were once covered by glaciers. In many mountains, the glacial ice carved deep valleys. The basins that were created filled with water. On the flatter land, glaciers gouged hollows in the land and deposited rocks and soil as they melted. Many lakes formed in these hollows and holes that were created by glacial deposits.

Lakes can form in regions where limestone underlies the soil. Underground water slowly dissolves the limestone rock. When the surface collapses, a sinkhole will form and fill with water forming beautiful crystal clear lakes.

Many artificial lakes have been formed by building dams across rivers to control the runoff of water. These dams are built for several reasons: to protect the surrounding area from floods, to provide a source of water for drinking and irrigation, to provide water to generate electricity, and for recreation.

The economic uses of lakes are numerous. Some of the larger lakes are used as travel and trade routes. Raw materials and other products are carried by boats across the lakes to industrial centers. Lakes provide an important source of water for irrigation. They are a natural reservoir of water for communities. Lakes created as storage reservoirs can be used to generate electric power. People use lakes for a variety of recreational activities such as swimming, boating, fishing, and water skiing.

Rivers

A river is a natural moving stream of fresh water. It is pulled downhill by gravity across the land. The flow of the river plays a role in the water cycle. With this cycle, water goes from the sea to the land and back again. Precipitation that falls on land evaporates back into the atmosphere, soaks into the ground, or runs off into rivers and streams and then into the ocean.

Like the oceans, rivers have been used for transportation for hundreds of years. Their waters provide routes for transporting people, their tools, food, and other necessities of life. Larger rivers are still used for economical transportation of goods and people.

Today the water from rivers is used in the cities for the people to drink. It is used by the farmers to irrigate crops. Because rivers often contain edible fish, they serve as a supply of food. Rivers are also used for recreational fishing, boating, water skiing, and swimming.

Icecaps and Glaciers

Snow that falls on land melts to form runoff. However, there are places on Earth where it is too cold for all the snow to melt. As the snow falls year after year, it piles up. The weight of the snow compacts it and it becomes ice. It is estimated that approximately 75 percent of all the freshwater on Earth is in the form of ice. The polar regions are almost completely covered by ice.

A glacier is a large mass of snow and ice that forms when the rate of accumulation exceeds the melt rate. Glaciers that cover large areas of land are called icecaps. Icecaps occur in places where the climate stays cold all year. An icecap must have land under it. Icecaps cover most of Antarctica and Greenland. At the North Pole there is no continent and the ice sinks and melts under pressure. At the South Pole there is a continent so ice has built up and created an icecap.

Edges of icecaps and glaciers break off when they reach the sea. These huge pieces of ice float away as icebergs.

The Earth's Features

Topic
Earth's Features

Key Question
What are Earth's major landforms?

Focus
The students will observe and compare their nearby environment with other physical features of the Earth through making models and drawing pictures.

Guiding Documents
NSE Standards
- *Ask a question about objects, organisms, and events in the environment.*
- *Communicate investigations and explanations.*

Project 2061 Benchmarks
- *A model of something is different from the real thing but can be used to learn something about the real thing.*
- *People can often learn about things around them by just observing those things carefully, but sometimes they can learn more by doing something to the things and noting what happens.*
- *Change is something that happens to many things.*

Science
Earth science
    geology
    physical features

Integrated Processes
Observing
Communicating
Comparing and contrasting
Collecting and recording data

Materials
Pictures of Earth's physical features (from magazines or travel agencies)
Globe showing physical features
Material for dioramas
Sand and dirt
Aluminum roasting pan (18" x 12") or flat cardboard cartons

Background Information
Most globes depict Earth as a smooth, round ball. Two-dimensional pictures taken from space also present a similar perspective, but when you *experience* the "real world," you realize that Earth is not smooth and level. It has a very interesting and varied landscape including mountains, valleys, plains, deserts, oceans, lakes, rivers, and icecaps.

Mountains are large landforms that rise conspicuously above the surrounding areas. *Hills* and mountains differ only in the relative height of the features. A *foothill* is a low hill near the foot of a mountain or mountain range.

Valleys are depressions in Earth's surface, usually among ranges of hills and mountains. Quite often a valley has been formed by the erosive action of a river. It can also be the depression between two uplifted mountain ranges. *Plains* are extensive areas of relatively level grasslands with few trees. The Great Plains of the United States covers thousands of square miles in the central part of the country.

A *desert* is an area that has very little water available for plants and animals to use. Living things in the desert have adaptations to make the best use of small amounts of water. The skies over deserts are usually cloudless, and there is a great difference in temperatures between day and night.

Over two-thirds of Earth is covered by *oceans*. An ocean is a great body of salty water. The waters of the oceans are never still, they are kept in constant motion by the heat of the sun and the wind. Water leaves the surface of the ocean by evaporation and is carried away in the clouds, only to return eventually in precipitation.

Lakes are inland bodies of water that are usually fresh water. They may form in any kind of hollow or basin on Earth's surface. Water may enter a lake in the form of rain and snow, or it may come from rivers and streams that drain the surrounding country and flow into the lake. Lakes can be formed by constructing a dam that blocks a river.

A *river* is a natural moving stream of fresh water that flows downhill across the land. Rivers are often very wide and deep. Smaller, shallower waterways may be called brooks or creeks.

At the North and South Poles, ice has frozen to form vast *icecaps*. In some areas the ice is 10,000 feet thick. Icebergs often break off the icecap and float out to the open sea.

Management
1. The child-like characters depicted on the *Features of the Earth* pictures can be referred to as *Science Buddies*. If desired, use them as though they are guides which narrate explanations of the different features and the activities that occur in the areas.
2. It is advisable that the teacher color the landscape pictures included in the lesson before presenting them to the students.

3. Collect pictures of different kinds of landscapes from travel agencies, magazines, calendars, and picture postcards. Earth's features studied in this lesson will be mountains, valleys, lakes, oceans, deserts, plains, rivers, and icecaps.

4. To make the pan of soil for the model of the landscape, follow the directions found in this activity for *Sand and Dirt Landscape Mixture*. (There are many different mediums that can be used to make a landscape. Do not feel limited to only this one.)

5. Gather some picture books to be made available for the students to use (see *Bibliography*). Put the students in cooperative pairs to do the dioramas and collect pictures for the features of the earth.

6. This is a two-part activity. *Part 1* explores the globe as a model of the Earth and identifies various landforms. *Part 2* involves each group of students in researching one landform and making a report to the class. An alternative would be that each feature can be done as a whole class activity.

Procedure

Part 1

1. Invite the students to shut their eyes and ask them the question: "What do you think of when I say the word *Earth*?" Discuss their ideas. Guide the students to describe Earth as a place where people, plants, and animals live.

2. Show the students a physical feature globe. Explain that this is a model of the Earth we live on.

3. Review the different colors on the globe. Ask if anyone can point out the United States. If necessary, indicate the U.S.

4. Ask the students where water can be found.

5. Discuss with the students that Earth is not smooth but consists of a variety of features such as mountains, hills, and valleys.

6. Show the students pictures of many different kinds of landscapes (beaches, deserts, mountains, plains, icecaps, etc.). Use the pictures on a bulletin board display so that the students can refer to them during the lesson.

7. Using the pictures in the activity, ask the students to describe how the land in each looks. Record their ideas by making a class word bank for each landform with a picture at the top and the students' words beneath. Make a connection to life science by discussing the plants and animals that live in each area and how they meet their needs of food, water, shelter, and space.

Part 2

8. Take the students outside and look around the playground to find miniature landscapes, such as valleys, streams, and lakes. Have them sit in a circle and look as far as they can in any direction—up,

down, or sideways. Inform them that this is all part of the Earth: the air, the water, the rocks and soils. Discuss the features that are near the school. Ask who has been to the ocean, mountains, lakes, etc. Have the students describe their experiences.

9. Discuss the dominate feature of your area so the students will be familiar with this landform. For example, if you live near the ocean, the ideal feature to start with is the ocean picture. Gather research and picture books on the subject of ocean. Ask the students to describe what the ocean looks, smells, feels, tastes, and sounds like. Ask them if it looks the same under the water as on the surface. Have them describe the plants and animals that live in and around the ocean. Ask how we use the ocean.

10. Repeat this with each of the landforms pictured in this activity until the students are familiar with each one.

11. Divide the students into cooperative pairs. Have each pair be responsible for one landform. Direct them to gather books to do research and make a report on their landform to present to the class. The students could draw a picture of their landform and glue sand, soil, twigs, or grass on appropriate areas. (The pictures could be used as a background for a diorama.) Encourage the students to draw people, plants, and animals to add to their landform picture.

12. Give prepared trays of wet sand and dirt to each group of students so they can build their own landscape. Encourage them to add landscaping features such as trees, rocks, grass, houses, rivers, lakes, and cacti as appropriate.

13. Have the students write a story about their landscape using words previously listed for each landscape.

14. Compare and contrast the various landforms as to rocks, plants and animals, climate, if people can live in the area, what type of work can be carried on in the landform.

15. Place two Earth feature pictures side-by-side and compare and contrast their characteristics.

16. Culminate the study by using the activities described in *Features of the Earth Games*. Encourage the students to draw pictures to make their own gameboard so they can review the names of each feature.

Discussion

1. Which of Earth's features do we have in or near our community?

2. Which picture shows an ocean?... mountain?... desert?... etc. Give some words you would use to describe an ocean... mountain... etc.

3. Describe the landscape you made. What details in it do you think are most important? Why?

4. Compare your landscape with that of another group. How are they alike?... different?

5. Why do you think it is important to know about the Earth's features?

8

6. Think about some of the plants and animals that live in your area. What are some of their special needs? How does the feature of the Earth that is nearby help these plants and animals meet their needs.

7. Which feature would you like to visit? Why?

8. Which feature of the Earth would you like to do research on and build a model?

Extensions

1. Make the *Layered Book of Earth's Features*. Instructions are provided. Be creative with what is put on the different sheets.
   Some suggestions:
   a. Put an additional white strip on first that says MY EARTH HAS...
   b. Instead of painting the *plains* page, fringe the top to resemble prairie grass.
   c. Assemble the pages and staple along the side so it can be read as book.
   d. Make the sentences cumulative on each page and underline the feature that applies to that page.

   e. Put appropriate animals on each page of the book.

2. Have the students imagine what it would be like to be on a high mountain. Have them describe what they could see, what the air would be like, what sounds they would hear. Ask them if there would be any trees, other vegetation, snow, etc.

Curriculum Correlation

Language Arts
Have the students write what it would be like to be in the middle of a lake in a sailboat or some other scenario that applies to a feature of the Earth.

Music
Play the selection "On the Trail" from the *Grand Canyon Suite* by Ferde Grofe.

Math
Make a class graph depicting the number of children who have actually visited the various features of Earth. Use the mathematical terms of *greater than, less than,* and *equal to* when comparing and contrasting the data.

Social Studies/Geography
Get a road map of your town or city. Mark, color, or cut and glue symbols to show where land and water features are found.

Home Link

The pages of the *Features of the Earth* book can be copied and stapled together for the students to take home and share with their parents.

Recipes for Earth's Features

Sand and Dirt Landscape Mixture
   To make a pan of soil for a model of the landscape, use three parts of sand to one part of dirt. Mix well and put enough water in the mixture so it can be molded. Mist the model with water every couple of days to keep the sand moist so the features remain firm. Put into a large pan or box so the feature can be moved around.

Play Clay
   1 1/2 cups flour
   1/2 cup salt
   1/2 cup water
   1/4 cup vegetable oil
   Food coloring

Mix the flour and salt. Slowly add the oil, water, and coloring. Knead the clay. If it is sticky, add more flour. Place in an air-tight plastic container and store in the refrigerator. Bake in the oven at 200° for two hours. The clay can be painted with poster paints or painted with white glue. When the glue dries, markers or water colors can be used.

Soft Clay
(classroom quantity)
   5 cups flour
   1 cup salt
   2 tablespoons alum
   2 tablespoons oil
   3 cups very hot water
   Food coloring

Stir all the ingredients together. Keep in the refrigerator to keep fresh.

Features of the Earth Games

   These games provide a way of reinforcing the concepts introduced in the activity *The Earth's Features*. The same pictures the students have become acquainted with in the activity are used in the games; hopefully students will apply what they learned from their own research and that of others. The *Word Board* and *Picture Board* pages are used for gameboards and playing cards. To make the playing cards, simply cut out the picture/word squares.

1. Copy onto card stock or tagboard a double-sided gameboard (picture on one side, words on other). Make one copy for every two students. For continued use, the gameboard should be laminated.

2. Make as many sheets of word squares as you have gameboards. Cut apart.

3. Make as many sheets of picture squares as you have game boards. Cut apart.

4. Make two sets of overhead picture and word squares so the game can be modeled on the overhead.

5. For the *Feature Concentration* game, copy double sets of picture cards, laminate, and cut apart.

Features of the Earth Book

Features of the Earth Book

Features of the Earth Book

Features of the Earth Book

13

Features of the Earth Book

Features of the Earth Book

15

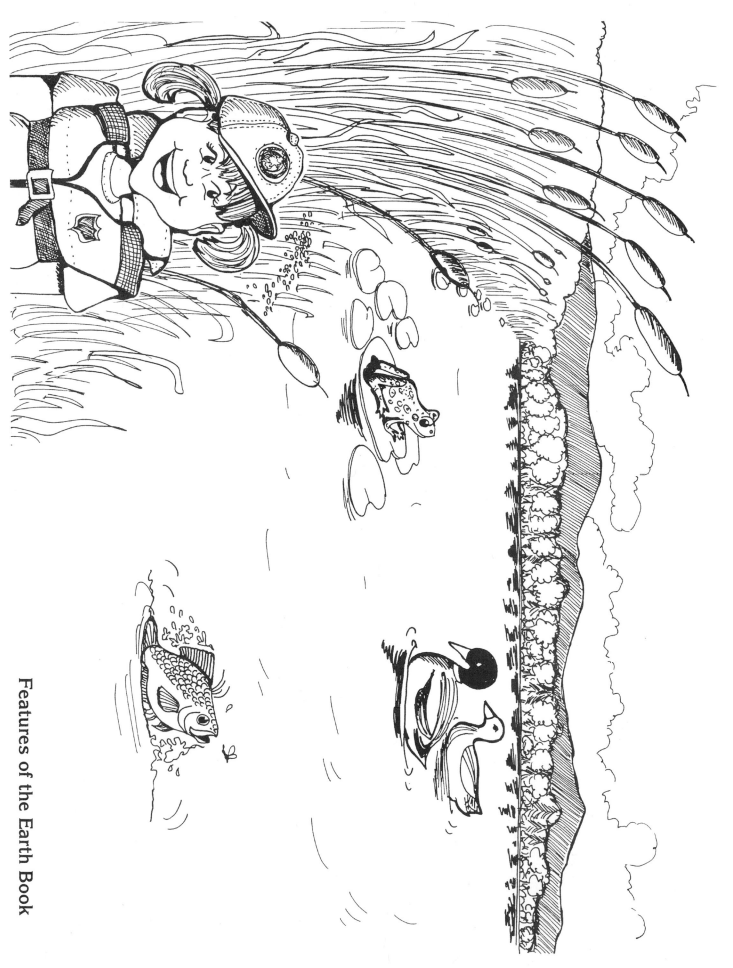

16

Features of the Earth Book

Features of the Earth Book

Game Cards
for
Features of the Earth

Game 1 — *Picture Lotto*

Objective: Match pictures of the features of the Earth

Procedure
   (Model the game on the overhead projector until the students become acquainted with the process.)
   1. Have students work in pairs.
   2. Ask them to turn the gameboard *picture-side* up.
   3. Pass out the picture cards.
   4. Tell the students to find the mountain on their gameboard and place the matching picture card on it.
   5. Continue until all the pictures have been covered.

Game 2 — *Picture Word Match*

Objective: Match the features of the Earth words to their corresponding pictures

Procedure
   (Model the game on the overhead projector until the students become acquainted with the process.)
   1. Have students work in pairs.
   2. Ask them to turn the gameboard *picture-side* up.
   3. Pass out the word cards.
   4. Tell the students to find the word *mountain* and put it on the picture of a mountain.
   5. Continue until all the pictures have been covered with their corresponding words.

Game 3 — *Word Picture Match*

Objective: Match the pictures of the features of the Earth to their corresponding words

Procedure
   (Model the game on the overhead projector until the students become acquainted with the process.)
   1. Have students work in pairs.
   2. Ask them to turn the gameboard *word-side* up.
   3. Pass out the picture cards.
   4. Tell the students to find the picture of the mountain and put it on the word *mountain*.
   5. Continue until all the words have been covered with their corresponding pictures.

Game 4 — *Feature Concentration*

Objective: Match two identical features of the Earth picture cards.

Procedure
   1. Have students work in pairs.
   2. Shuffle a double set of cards and randomly place them face down.
   3. Tell the students to turn over any two cards trying to find two identical pictures. If they are successful, have them keep the pair of cards; if not, direct them to turn the cards back over. When two cards are matched, have the student describe the feature with a sentence.
   4. Tell them that the trick is to remember where the cards are located so on their next turn they can match two identical cards.
   5. Have them continue play until all the cards have been picked up.

Features of the Earth

plains	lakes	mountains
deserts	Earth	ice caps
Oceans	rivers	valleys

Picture Board

21

Layered Book of Earth's Features

You will need: 9"x12" construction paper
sand, glue
scissors
paint

cut to

dark blue 3"   white 11"
light brown 4"   sky blue 12"
manila 5½"
green 6½"
dark brown 9½"

Do this:

3" — My Earth has oceans.
9"

dark blue sponge paint green flecks

4" — My Earth has plains.

light brown streak green paint for grass

white

11" — My Earth has ice caps.

5½" — My Earth has deserts

manila glue sand

12" — My Earth has ice caps, mountains, rivers, lakes, valleys, deserts plains, oceans, and me!

sky blue pasted or painted portrait

6½" — My Earth has valleys.

green fringe top for grass

9½" — My Earth has mountains, rivers and lakes.
9"

Dark brown

blue pasted on shape

Put it all together and staple

My Earth has oceans.

What does my Earth have?

My Earth has oceans.

My Earth has plains.

My Earth has deserts.

My Earth has valleys.

My Earth has mountains, rivers, and lakes.

My Earth has ice caps.

My Earth has ice caps, mountains, rivers, lakes, valleys, deserts, plains, oceans, and me!

Cut out and use for the layered book for Features of the Earth.

My Rock

Topic
Rocks

Key Question
How would you describe your rock?

Focus
By making careful observations, students will describe various properties of their rocks.

Guiding Documents
NCTM Standards
- *Make and use measurements in problems and everyday situations*

NSE Standards
- *Objects have many observable properties, including size, weight, shape, color, temperature, and the ability to react with other substances. Those properties can be measured using tools, such as rulers, balances, and thermometers.*
- *Employ simple equipment and tools to gather data and extend the senses.*
- *Simple instruments, such as magnifiers, thermometers, and rulers, provide more information than scientists obtain using only their senses.*

Project 2061 Benchmarks
- *Chunks of rocks come in many sizes and shapes, from boulders to grains of sand and even smaller.*
- *Describing things as accurately as possible is important in science because it enables people to compare their observations with those of others.*
- *Objects can be described in terms of the materials they are made of (clay, cloth, paper, etc.) and their physical properties (color, size, shape, weight, texture, flexibility, etc.)*
- *Often a person can find out about a group of things by studying just a few of them.*

Math
Measurement
   length, width
   mass

Science
Earth science
   rocks

Integrated Processes
Observing
Collecting and recording data
Comparing and contrasting

Classifying
Predicting
Communicating

Materials
For Part 1:
   rock
   string
   glue
   balance
   Teddy Bear Counters

For Part 2:
   rock
   measuring tape
   balance
   masses
   magnifying lens
   container with water

Background Information
   Rocks shape much of our Earth. They form the mountains and hills. We do not see many of the rocks because they are covered with soil and vegetation. Pebbles, soil, and sand all have small pieces of eroded rock in them.
   Rocks consist of one or more minerals. The minerals give color, hardness, and sparkle to rocks.

Management
1. Encourage each student to bring a special rock from home. Emphasize that the rock should be a size that can be carried in one hand, preferably the size of an egg.
2. Have extra rocks for students who don't bring any or for those who bring unsuitable rocks.
3. Set up the balances and have Teddy Bear Counters available.
4. This activity is divided into two parts. *Part 1* is for younger students while *Part 2, I Found a Rock,* is for older students.

Procedure
Part 1
1. Using the interesting and unusual rock specimens that the students have found and brought to school, ask the students to observe their rocks carefully. Ask them to look at other rocks within their group and tell how their rock is alike and different. Have students share what they discovered.
2. In the sharing time, discuss what properties they observed: color, texture, size, etc. Tell students that they will be making a record of the properties of their

rock, but first they need to draw and color a picture of their rock on their activity sheet and give it a name.

3. Direct them to determine and write whether their rock is light or dark, rough or smooth. Discussion may arise as to what is considered "light" and "dark" or "rough" and "smooth." Because these terms are relative, let students determine this classification within small groups.

4. Brainstorm what colors they might be looking for in a rock. Ask students to describe their rock using color words. For example: "My rock is mostly black with white freckles in it." "My rock is light brown with a big dark brown spot."

5. Tell the students that they will make some more observations of their rocks, but this time they will observe using measurements. Ask them to share ways in which they could measure their rocks. After acknowledging some responses, tell them that they will use string to measure how long and big around their rocks are. (Assure students that they will be given time to use their measurement strategies.) Demonstrate the method and direct the students to glue their strings to their activity sheet.

6. Next, ask them to estimate how many Teddy Bear Counters it will take to balance their rock. Have them find the actual count of Teddy Bears needed to balance the rock and record this amount on the activity sheet.

7. If appropriate, use *My Rock's Story* for the students to write a creative story about their rock.

Part 2

1. Ask each student to choose a favorite rock, examine it, and become very familiar with it.

2. Direct the students to use the recording sheet *I Found A Rock* to record some of the observations they made of their rock and describe what makes it special.

3. Encourage students to use a magnifying lens to examine the texture of their rock. Direct them to look at the size of the mineral grains in their rock and compare the grain size with others in their group. Discuss that large grains or chunks represent a coarse texture while small grains represent a fine texture.

4. Distribute the containers with water. Have the students submerge their rocks in the water. Ask them to describe their rock when it is wet.

5. Have students measure the circumference of the rock and find its mass.

6. Ask them to find another rock that is similar to theirs and explain the ways the two rocks are alike. Urge them to try to find at least four ways that the rocks are the same and four ways they are different. Encourage them to record the similarities and differences.

7. Direct students to put their rocks in a large group. Have them trade recording sheets to see if others can identify the rocks.

Discussion

1. How are the rocks alike?
2. What was the heaviest rock? How many Teddy Bear Counters did it take to balance your rock?
3. Was the heaviest rock also the biggest? How can you prove it?
4. How many different colors were found in all the rocks? (List them.)
5. Which rocks were one solid color?
6. Who had the longest rock? Was it also the biggest around?
7. What makes your rock different from everybody else's?
8. What are some other things you would like to know about your rock? Think of some ways you could find out.
9. How does water make the rocks look different?

Extensions

1. Discuss with the students all the words that can be used to classify rocks. (big, small, long, heavy, etc.) Use *Show Me a Rock* to question the students about their rocks.
2. Allow time for students to use their measurement strategies from *Part 1*.

Curriculum Correlation

Language Arts

Put several rocks on the table. Have the students observe the rocks carefully. Then on a piece of paper, have them write some words that describe one of the rocks. Exchange the papers with another student and see if the partner can pick out the rock that was described.

Math

For experience using the balance, encourage the students to combine their rock, or rocks, and Teddy Bear Counters to balance someone else's rock.

Home Link

Suggest to the parents that the students collect rocks around the home and sort them into a collection.

My Rock

Geologist_____

This is a picture of my rock. I drew it.

My rock's name is_____.

My rock is_____.(light or dark)

My rock is_____.(rough or smooth)

The colors in my rock are_____

_____.

My rock is special because_____

_____.

My Rock's Measurements

Name _____

My rock is this long.
(glue string here)

My rock is this big around.
(glue string here)

My rock measures this many inches.

Lay your rock on the ruler. Mark its length. Color it in.

| 1 | 2 | 3 | 4 | 5 | 6 | 7 | 8 | 9 |

Balancing with Bears

Name _____

I think ▢ bears will balance my rock.

I found that ▢ bears balanced my rock.

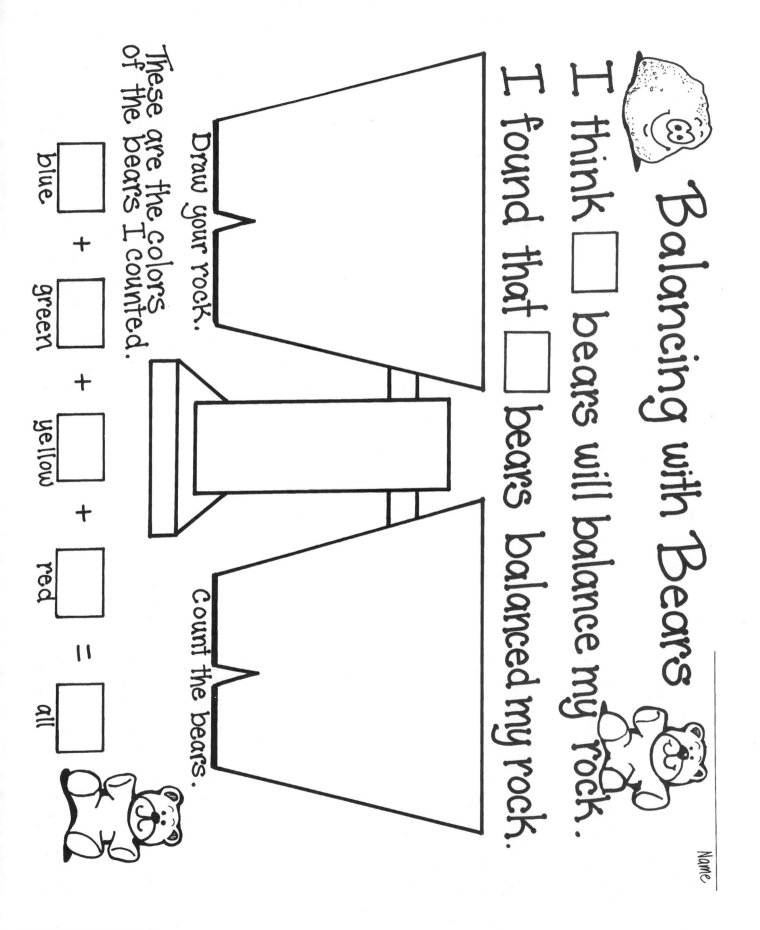

Draw your rock.

Count the bears.

These are the colors
of the bears I counted.

▢ + ▢ + ▢ + ▢ = ▢
blue green yellow red all

My Rock's Story

_____

_____

_____

_____

_____

_____

_____

_____

_____

_____

_____

_____

written by:_____

I Found a Rock

1. My rock is special because. _____

_____

_____

2. This is a picture of my rock. _____

3. This is a picture of my rock wet. _____

4. My rock has these colors_____.

5. I touch my rock and it feels_____.

6. I measure my rock.

length = _____ long

height = _____ high

7. I found the mass of my rock.

= _____

Show Me a Rock...

1. Show me a big rock.
2. Show me a small rock.
3. Show me a black rock. (white, grey, red, yellow)
4. Show me a tiny rock.
5. Show me a rock with stripes.
6. Show me a long rock.
7. Show me a flat rock.
8. Show me a round rock.
9. Show me a shiny rock.
10. Show me a rough rock.
11. Show me a smooth rock.
12. Show me a heavy rock.
13. Show me a sparkly rock.
14. Show me a bumpy rock.
15. Show me a speckled rock.
16. Show me a rock with lines.
17. Show me a rock shaped like an egg.

Topic
Properties of rocks

Key Question
How can a rock be described?

Focus
The students will observe the physical properties of rocks and group them according to certain attributes.

Guiding Documents
NCTM Standards
- *Recognize, describe, extend, and create variety of patterns*
- *Collect, organize, and describe data*

NSE Standards
- *Objects have many observable properties, including size, weight, shape, color, temperature, and the ability to react with other substances. Those properties can be measured using tools, such as rulers, balances, and thermometers.*
- *Employ simple equipment and tools to gather data and extend the senses.*
- *Simple instruments, such as magnifiers, thermometers, and rulers, provide more information than scientists obtain using only their senses.*

Project 2061 Benchmarks
- *Objects can be described in terms of the materials they are made of (clay, cloth, paper, etc.) and their physical properties (color, size, shape, weight, texture, flexibility, etc.).*
- *Chunks of rock come in many sizes and different shapes, from boulders to grains of sand and even smaller.*
- *Numbers can be used to count things, place them in order, or name them.*
- *Simple graphs can help to tell about observations.*
- *Often a person can find out about a groups of things by studying just a few of them.*
- *One way to describe something is to say how it is like something else.*
- *Things in nature and things people make have very different sizes, weights, ages and speeds.*

Math
Measurement
   length
   mass
Graphing

Science
Earth science
   rocks

Integrated Processes
Observing
Classifying
Comparing and contrasting
Collecting and recording data
Interpreting data
Generalizing

Materials
Rocks (small enough to fit in one hand)
Paper, 9" x 12"
Magnifying lenses
Butcher paper
Tape measure
Balances (optional)
Masses, non-customary or customary (optional)
Colored self-stick dots
Note cards, 3" x 5" (optional)
Chart paper

Background Information
Rocks are fascinating to children and adults alike. They like to pick them up and hold them, feel, rub, and examine them. As rocks are collected, the students will find that they do not all look and feel the same. Some of the rocks are smooth and shiny, while others are rough and dull. Some break along definite planes while others crumble. Rocks vary greatly in hardness, which is a convenient trait to use to identify and classify rocks. Studying rocks can lead to interesting "discoveries" which might include the number and sizes of holes in a rock, the different colors of the minerals in the rock, or the different materials making up the rock. The students might find fossils in sedimentary rocks.

Rocks can be found in many places; the crust of the Earth is mostly solid rock called bedrock. However, this rock is often covered by a thick layer of vegetation, sand, soil, gravel, or water. Occasionally, the solid rock is visible through the soil as outcrops. Erosion breaks up the solid rock into smaller pieces and eventually into soil.

Management
1. These activities should be done over a period of several days.
2. They are best done in small groups of two or four students. Allow an extended block of time for exploration and investigation.

3. Have the students collect rocks from around the school or around their neighborhood. Display them for a few days giving the students a chance to touch and talk about the rocks.

4. If possible take the students on a field trip to hunt rocks for a collection so they will have an opportunity to see where rocks are located.

5. Have extra rocks available if students do not bring in rocks from their home.

6. Try to get rocks with as many different properties as possible: shiny, dull, rough, smooth, dark, layered, banded, one color, and many colors.

7. You may find interesting rocks at building sites, road construction sites, beaches, or rivers. In those cases where there just aren't any rocks to be found, you may be forced to use landscaping rocks or rocks from rock collectors. If enough time is given for the task, students may write to relatives living in areas where there are rocks to request that they send one or two.

Procedure
Part 1

1. Tell the students to find and bring in rocks that are no bigger than a fist.

2. Display the class collection of rocks and allow the students to observe and explore their properties. Let students work with a partner to discuss, describe, and hold the collection before doing the more formal activity of classifying and grouping.

3. After the students have been able to explore the collection of rocks, hold up two rocks and have the whole class participate in a discussion of observations they can make. List the words they use on a chart. If the students have difficulty in thinking up words, suggest some that relate to size, shape, color, texture, hardness, or heaviness.

4. Tell the students that they are teams of geologists who are going to study some rocks and group together those rocks that are alike in some way. Let each group choose eight rocks, and then develop their own classification schemes for the rocks, such as flat, smooth, dark, or odd-shaped. (Don't worry about the "correct" characteristics at this time.)

5. Have the students group them on a 9" x 12" sheet of paper. When they are finished, let each group explain how they grouped the rocks.

Discussion

1. Explain how you sorted your rocks. What properties did you use?

2. What are some of the words you used to describe your rock groups?

3. How were your rules for sorting the same as your neighboring team's? How were they different?

4. If you were going to sort these same rocks again, what are some other rules you could use? If you were to sort your neighboring team's rocks, what rules could you use?

5. Where did your team find most of your rocks? Where else could you look?

6. Why do you think it is important to be able to sort rocks?

7. What did you learn about rocks today?

Procedure
Part 2

1. Gather the whole class together and hold up two rocks. Ask the students to describe the two rocks. Record these observations. (Observations might be related to size, shape, color, texture, luster, hardness, or heaviness.)

2. Explain to the students that they are going to make a chart that tells about the rocks. Enlarge (3' x 5') the first activity sheet to make a class matrix on chart paper. Label the left vertical column *Our Rocks*. Ask the class to name some of the properties of the rocks. Use these suggestions (for example: shiny, smooth, rough, flat) as labels in the spaces across the top.

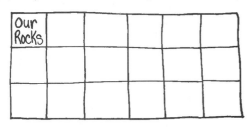

3. Place a red sticky dot on the first rock and a blue dot on the second. Hold up the first rock. Ask the class to decide which attributes describe this rock. Make an x in the box of each of the attributes. (Example: The rock with a red sticky dot is shiny = x, dark = x, and round = x. It is not chalky or heavy so no "x" will be placed in those columns.) Place this rock in the first space under Our Rocks and next to the "x" marks on the chart. Proceed to the next rock and model how to develop a description for it.

Our Rocks	shiny	dark	round	chalky	heavy
●	X	X	X		
▢				X	X

4. During this modeling process, students should begin to see that we need to come up with a way of explaining certain terms such as "heavy" and "big." Ask them how they would determine "heavy." If necessary, suggest that they could use a balance and determine a number of Teddy Bear

Counters or gram masses to be "heavy." (For example: All rocks with a mass of ten or more Teddy Bear Counters will be considered "heavy." All those rocks with a mass of nine or less Teddy Bear Counters will be considered "light.") This should be a class decision. Place these definitions in the legend at the bottom of the chart so others will understand them.

Heavy-10 or more
Light- 9 or less

5. Allow the students to determine what they would consider a *small, medium,* and *large* rock to be. For example: with 3" x 5" note cards, cut a 2 cm square from the middle of one, a 3 cm square from a second card, and a 4 cm square from the middle of a third card. Label these cards *small, medium,* and *large,* according to the size of the hole. Have the students try to slip their rocks through these holes. If the rocks slip through the 2 cm square hole, they would be classified as a small rocks. (Be careful of long narrow rocks!) If they don't easily slip through, have students try the next size larger, and so forth. The record for whichever method is chosen should be placed in a legend on the chart.

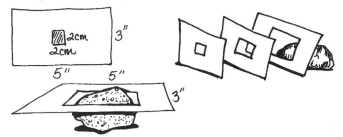

6. Once the students have learned how to determine if their rocks fit the properties defined, tell them to choose one rock and determine how it would be placed on the large class chart.

Discussion
1. Was your rock large, medium, or small? How could you tell?
2. Was your rock heavy or light? How do you know?
3. How many teams had heavy rocks?
4. How many teams had small rocks?
5. Describe your rock to the class.
6. Use the chart to try to identify a rock that is described on it. Explain how your rock is different than the rock you identified from the chart. How is it the same?

Procedure
Part 3
1. Place the students in teams of four and have them pick four different rocks.
2. Allow time for all the team members to study their four rocks. Give each team a piece of large chart paper and tell them to make a matrix like the one previously used to describe rocks. Direct them to

write *Our Rocks* in the upper left-hand column. Then ask them to write descriptive words across the top of the other columns. Urge them to use words that would describe properties they observed in their four rocks. (Remind them that each rock does not need to have each property listed; they will merely not put an x in that box.)

3. Distribute eight sticky dots, two each of a color, to each group. Have the students place a different colored sticky dot on each rock to designate the different rocks. Direct them to place one rock in each box under the label *Our Rocks.* Have them take the same colored sticky dot as the rock and stick it onto the chart next to the rock.

4. Allow time for the students to look carefully for the properties of each rock. Direct them to take the first rock and check it for each property listed. For each property that is observed in that rock, have them place an x in the box in that row under that property. Remind them, if necessary, that some rocks may have more than one property. (Example: The first rock is heavy = x, round = x, and sparkly = x. It is not flaky or dark so no x will be placed in those columns.)

5. Repeat the process with the other three rocks.

6. Now the fun begins! Tell the students to come up with a name for themselves as a team of geologists and put it on their chart paper. Direct them to fold the left side of their chart under to hide the colored sticky dots and to place their rocks to the side in a group.

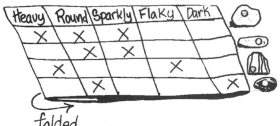

7. Allow teams to change tables and to look at the chart and rock samples of another team. Direct them to place the rocks on the right side of the chart according to the properties checked. When they are finished, have them fold out the flap to reveal the team's original charting. They can then compare their results with the original team's charting. Once they have made the comparisons, have them fold under the flap put the rocks in a pile and move to another team's rocks and chart and repeat the procedure.

8. After the class has had a chance to try several "geologists' charting systems," bring the whole class together and discuss what made it easy, or difficult to determine a certain rock.

Discussion

1. Did you find it easy or difficult to match the rocks with the charting that other teams made? Explain.
2. What were some of the rules that were used by most of the geologist teams in the room? Why do you think so many of us used them? How many teams used _____ rule?
3. What were some rules you found that were not used very often? Why do you think this happened? How many more teams used the _____ (most frequent rule) than used _____ (least frequent rule)? Name two other rules that were used more often than _____ (least frequent rule).
4. If you were going to do this lesson again with a different set of rocks, how would you set up your new chart? What would you do that is the same as this one? What would you change?
5. What did you enjoy about this activity?

Part 4

1. Depending on the understanding and grade level of the students, give the teams an open-ended, blank 5" x 5" chart to use as an evaluation of this process. Have students choose a new set of rocks. Direct them to add property labels, realistically illustrate the rocks, and repeat the procedure of marking the columns as they did earlier.

Curriculum Correlation

Literature

Gans, Roma. *Rock Collecting*. Harper & Row. NY. 1984.

Home Link

Encourage the students to make a rock collection at home. Tell them to separate them as they have in class and make a rock chart matrix. Suggest to them that they use the matrix to play a *Guess Which Rock I Am* game with their families.

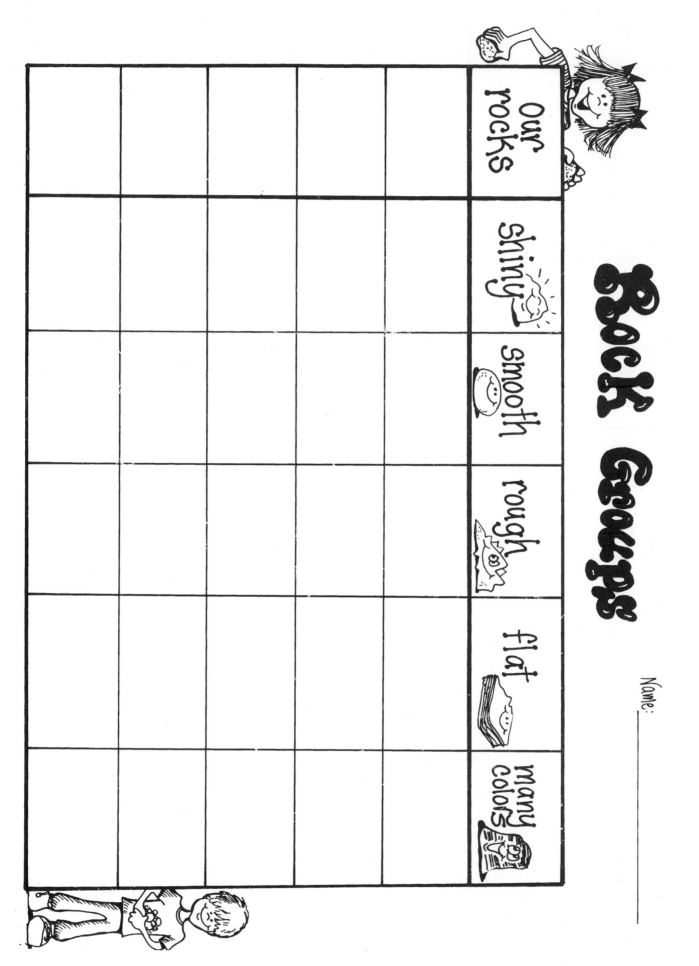

Rock Groups

Name: _____

Our rocks	shiny	smooth	rough	flat	many colors

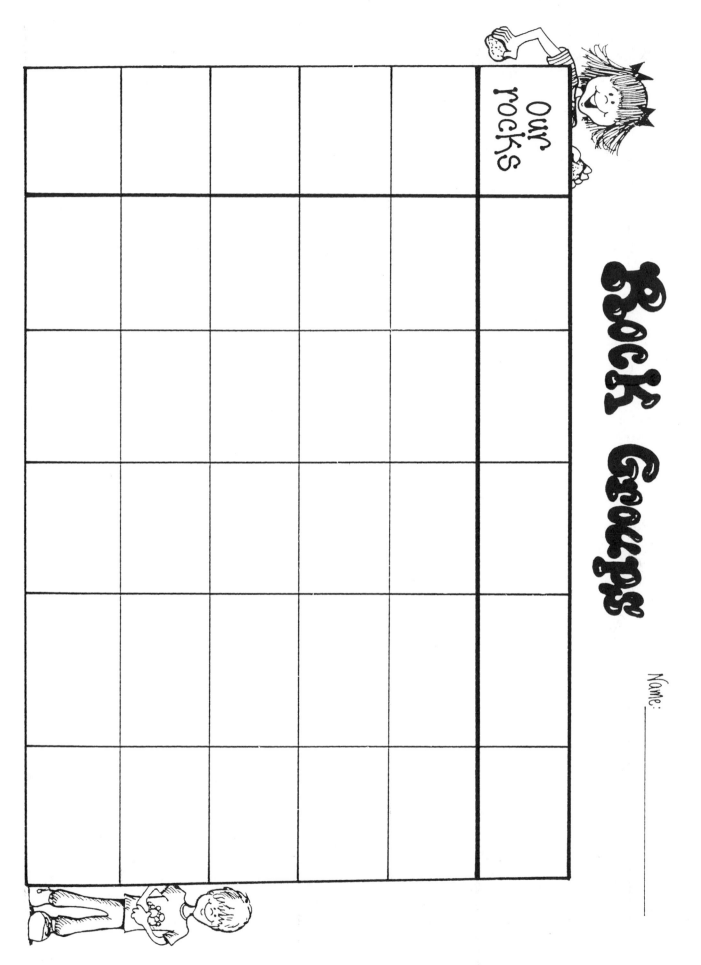

Rock Groups

Our rocks

Name: _____

Rocks and More Rocks

Topic
Properties of rocks

Key Question
How can we sort and group rocks?

Focus
The students will observe the physical properties of rocks and group them accordingly.

Guiding Documents
NCTM Standards
- *Recognize, describe, extend, and create a variety of patterns*
- *Collect, organize, and describe data*

NSE Standards
- *Objects have many observable properties, including size, weight, shape, color, temperature, and the ability to react with other substances. Those properties can be measured using tools, such as rulers, balances, and thermometers.*
- *Employ simple equipment and tools to gather data and extend the senses.*
- *Simple instruments, such as magnifiers, thermometers, and rulers, provide more information than scientists obtain using only their senses.*

Project 2061 Benchmarks
- *Objects can be described in terms of the materials they are made of (clay, cloth, paper, etc.) and their physical properties (color, size, shape, weight, texture, flexibility, etc.).*
- *Chunks of rocks come in many different sizes and shapes, from boulders to grains of sand and even smaller.*

Math
Measuring
   length
   mass
Sequencing
Binary classification
Venn diagram

Science
Earth science
   rocks

Integrated Processes
Observing
Classifying

Comparing and contrasting
Collecting and recording data
Generalizing

Materials
Rocks (see *Management*)
Butcher paper
Magnifying lens
Tape measure
Yarn

Background Information
History records that rocks have been used as tools, as weapons, and as building materials. The Stone Age in history is so named because crude, fist-sized tools such as spears, cutting tools, diggers, and chippers were made from rocks.

A rock is defined by geologists as a substance that is composed of one or more minerals. A mineral is similar in chemical composition and structure throughout all its parts. Rocks are made up of minerals, but rocks themselves are not minerals. The tiny specks of color in rocks are the different minerals. Rocks have certain properties by which they may be described and identified. They vary in color, hardness, texture, chemical makeup, weight, and are shiny or dull.

Rocks have many properties that can be used to sort them. Some are color, shape, luster, texture, weight, and size. By having young learners classify rocks according to properties that are familiar to them, they are developing their observational skills as well as their appreciation for the beauty and infinite variety of rocks in their environment.

Management
1. Use the same rocks that the students have used in *Rock Groups*.
2. This activity focuses on three organizational methods for recording properties of rocks: binary classification, sequencing, and Venn diagramming. It also presents a *Challenge* for students to make chains of rocks according to different properties (see *Part 4*).

Procedure
Part 1
Binary Classification
1. This activity can be done with the whole class and led by the teacher. Using a large piece of butcher paper or chart paper, choose about 16 rock samples. Place all the rocks in a pile at the top of the paper. Draw a circle around the pile.

2. Choose one property, such as large. Move all the large rocks to a separate pile, then make another pile of the remaining rocks. Draw a circle around each pile and label the one *large* and the other *not large.*

3. Look at the large rocks, choose a physical property that will allow the rocks to again be divided. Draw a circle around each pile and label.

4. Repeat with the other group of rocks. Be sure to include labels.

5. Keep dividing the piles using other physical properties. (Some ideas are: rough, not rough; heavy, not heavy; crumbly, not crumbly; one color, not just one color; shiny, not shiny.) Do this until each rock is by itself.

Part 2
Sequencing

1. To model the sequencing procedure, enlarge the following rock train diagram.

2. Direct the students to look at some variations of specific properties of the rocks. For example, there are some rocks that are light colored, others that are a little darker, and still others that are very dark. Discuss with the students what they have observed.

3. The teacher should take three rocks and ask students to help seriate them from lightest in color to darkest in color.

4. Repeat this procedure using another property such as size and seriate the three rocks from largest to smallest.

5. Practice a similar procedure using four other rocks.

6. Distribute the *Rock Train* sheet and direct the students to get rocks from the large collection. Encourage them to seriate five rocks according to the attribute on the engine of the train.

Part 3
Venn Diagramming

1. Take a collection of six rocks that have several attributes.

2. Put two rings of yarn on the table. Spread out the rocks so students can observe them. Together come up with two rules for sorting the rocks (shiny and large, for example).

3. Write one of the rules on a small piece of paper and put it inside one circle. Have the students put the rocks that follow that rule in the circle. When finished, write the other rule on another small piece of paper and place in the second circle. Place rocks that have that property inside the second circle.

4. Ask students if they noticed anything unusual about the rocks that went in the circles. If they don't come up with it on their own, guide them to determine that some rocks fit both rules and should be placed in both circles. Ask them how they could show that using the two circles of yarn. If necessary, begin to move one of the circles to overlap the other. Spend some time letting students understand that the intersection is a part of both circles. When students show an understanding, have them place those rocks with both properties in the intersection of the two circles.

5. Change one of the properties and challenge students to again sort the rocks.

Part 4
Challenges

1. If appropriate, challenge the students to make a chain of rocks so that each rock in line is different in one way from the neighbor rocks.

2. Present a new challenge: Make a chain of rocks so that each rock is different in two ways from the neighbor rock. Let others guess the rules.

3. Put the *Station Cards* and equipment at various stations around the room and challenge the students to complete them. They can record their results by recording
   a. how many tries they needed to find their rock,
   b. how many ways they were able to sort their rocks,
   c. how many cubes they needed to balance each rock? (Record results on *Balance Rocks* sheet), and
   d. the rule they used to put the rocks in order.

Discussion

1. Do you think you can find two rocks that are exactly alike? Explain. How are the rocks in the collection alike? How are they different?

2. Of the ways we have learned to sort rocks, which way did you like best? Why? Which way did you like the least?

3. When you sorted rocks, what were the properties you used?

4. Do you think all rocks are smooth? Explain. What do you think may have made some of them smooth? If you were going to look for smooth stones, where do you think you might look? (Use similar questions to explore different properties.)

5. What did you learn about rocks that you didn't know before you did this activity?

6. Think about all the rocks our class has been observing. Describe the kind of rock that you found most. Describe a rock of which you found only one.

7. Why do you think more _____ rocks were found than any other kind?

8. Which is the most beautiful rock in our collection? Why?

Extensions

1. Have students use their imaginations to make rock creations. For example, they could make a paperweight; a "rock band" group; a pebble person, insect, animal, or other creature; a rock village; etc.

2. Ask a rockhound to come to the classroom and share his/her collection of rocks and experiences in searching for rocks.

3. Make some small rock gardens with sand and stones. Use pictures of Japanese rock gardens to get some ideas.

4. Go on a scavenger hunt to find different kinds of rocks. The list could include a shiny rock, a smooth rock, a tiny rock.

Curriculum Correlation

Art

1. Make collages with colored aquarium gravel. The students can paint their rocks or add other art materials such as paper, material, or sticks to make a rock sculpture.

2. Look at one rock and try to see if it or any part of it looks like a face, an animal, or something else that is familiar. Describe this to someone else. Draw it.

Language Arts

Make a rock-shaped book and have the students write a story about their rock and illustrate it.

Math

Fill a jar with rocks and have the students estimate the total number of rocks. As a class count them by 2's, 5's, 10's on the overhead projector.

Social Studies

1. Visit a rock and gem store or that section of a museum near you.

2. Visit a gravel pit or a quarry where stone is taken from the Earth. Talk about the kind of stone that is found at each place.

Home Link

Encourage the students to make a rock collection at home. Tell them to separate them as they have in class, then explain their classifications to their parents. Have them show their parents their favorite rock and explain to their parents why it is special.

ROCK TRAIN

Name: _____

Roughest
Smoothest
Texture

Size
Largest
Smallest

Heaviest
Lightest
Mass

41

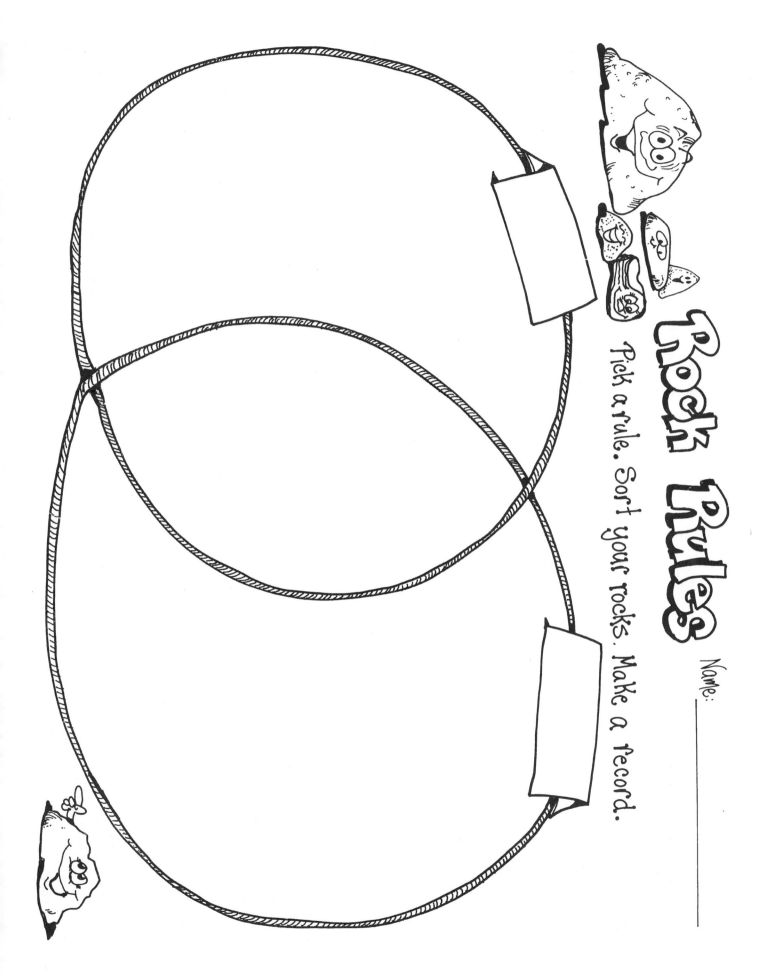

Rock Rules

Pick a rule. Sort your rocks. Make a record.

Name: _____

Find Your Rock

1. Look at your rock.

2. Put your rock in the bag.

3. Shut your eyes.

4. Find your rock in the bag.

Station Card

Sort Rocks

1. Look at the rocks.

2. Sort your rocks on the mat.

3. Try another mat.

Station Card

Balance Rocks

1. Put a rock on a balance.

2. Put cubes in to make the buckets even.

3. Count the cubes.

4. Try another rock.

Rock Train

1. Look at rocks.

2. Pick a rule.

3. Put rocks in order.

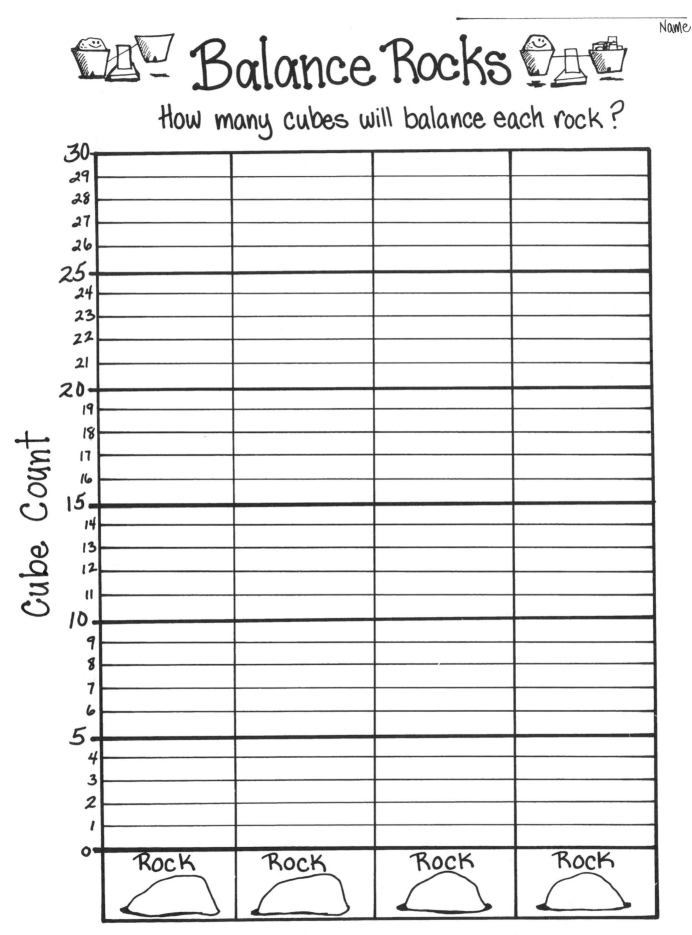

Balance Rocks

How many cubes will balance each rock?

Cube Count

	30				
	29				
	28				
	27				
	26				
	25				
	24				
	23				
	22				
	21				
	20				
	19				
	18				
	17				
	16				
	15				
	14				
	13				
	12				
	11				
	10				
	9				
	8				
	7				
	6				
	5				
	4				
	3				
	2				
	1				
	0	Rock	Rock	Rock	Rock

Name

Ice Breakers

Topic
Weathering of rock

Key Question
How can the freezing of water change rocks?

Focus
Primary students will investigate the effects of the freezing of water in simulated model rocks.

Guiding Documents
NCTM Standards
- *Systematically collect, organize, and describe data.*
- *Construct, read, and interpret tables, charts, and graphs.*

NSE Standards
- *Materials can exist in different states—solid, liquid, and gas. Some common materials, such as water, can be changed from one state to another by heating and cooling.*
- *The surface of the earth changes. Some changes are due to slow processes, such as erosion and weathering, and some changes are due to rapid process, such as landslides, volcanic eruptions, and earthquakes.*

Project 2061 Benchmarks
- *People can often learn about things around them by just observing those things carefully, but sometimes they can learn more by doing something to the things and noting what happens.*
- *Change is something that happens to many things.*
- *A model of something is different from the real thing but can be used to learn something about the real thing.*

Math
Graphing
Equalities and inequalities
Counting

Science
Earth science
    weathering
Physical science
    states of matter

Integrated Processes
Observing
Predicting
Collecting and recording data
Interpreting data
Applying
Generalizing
Communicating

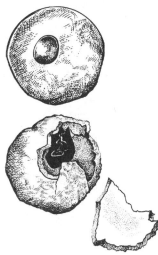

Materials
For the class:
    optional: *The Big Rock* by Bruce Hiscock. Macmillan Publishing Co. NY. 1988.
    bucket of water
    unglazed brick
For each group of four students:
    small water balloon
    1/2 cup of salt/flour clay or Plaster of Paris (see *Management*)
For each student:
    science journal

Background Information
Weathering is the set of processes that disintegrates or decomposes rocks (or other materials) at or near the surface of the Earth. It is the response of Earth's materials to a changing environment. The two general types of weathering are chemical and mechanical. They usually occur very slowly and often without our notice.

Mechanical weathering processes do not change the composition of rocks and other materials as do chemical weathering processes. Rather, *mechanical weathering* changes the size of the materials by breaking them into smaller pieces. Frost wedging, the mechanical weathering process investigated in this activity, is caused by the alternate freezing and thawing of water which seeps into the cracks of rocks. Because water expands about 9% when it freezes, it exerts a tremendous outward force.

To simulate a rock, a water-filled balloon is covered with salt/flour clay or Plaster of Paris. After the clay or Plaster of Paris has dried, the "rock" is put into a freezer overnight. The expansion of the water as it freezes will cause the water in the balloon to expand and break the clay or Plaster of Paris. Connections to weathering are made to show causes of changes in our Earth over time. The students also observe a change in the state of matter when the water changes from a liquid state to a solid state through the process of freezing.

In this activity, a second representation of ice breaking rocks is provided. In this example, an unglazed brick is soaked in a bucket of water for a couple of days prior to the activity. The wet brick will be placed into the freezer at the same time as the clay/plaster "rocks." The brick should show breakage after the water has frozen.

Management
1. Decide whether you want to use the salt/flour clay or Plaster of Paris to coat the outside of each balloon.

Recipe for salt/flour clay

1 cup flour
1/2 cup salt
1/3+ cup water (add water until clay is dough-like)
Mix ingredients together. Knead until smooth.

---

2. If using Plaster of Paris, follow mixing directions found on the package. Be cautious when mixing the powder so no one breathes the dust. You will want to use a thick clay-like mixture that students can handle.
3. Make arrangements to use a freezer at school.
4. Copy one science journal for each student.
5. Prior to the activity, fill the water balloons until they are about the size of golf balls. Tie the balloons so they won't leak.
6. If desired, a chunk of limestone or sandstone can be used instead of the brick. Make sure to place the rock or brick in the bucket of water two days prior to doing the activity.

Procedure

1. If available, read *The Big Rock* to the students. Discuss weathering.
2. Tell the students they are going to make a model of a rock so they can see what happens to rocks as water freezes.
3. Give each student or group a water balloon.
4. Using either the pre-mixed salt/flour clay or Plaster of Paris, tell the students to completely cover their balloons to make them resemble rocks. Make certain the covering is not excessively thick as this prolongs the necessary drying time.
5. Allow the rock models to thoroughly dry.
6. Ask students how their model rocks are like real rocks and how they are different.
7. Once the rock models are dry, place them and the brick into a freezer.
8. Mark the chart as to *prediction* or *result* by coloring in the thought bubbles at the top of the recording sheet *Will Your Model Rock Break When Frozen?*
9. The next day check the rocks and graph the actual results as to which rocks were broken and which ones were not.
10. Have students draw pictures in their science journals of what happens to rocks when water seeps into their cracks and freezes.

Discussion

1. What things in nature do you think make rocks break? [falling, earthquakes] Explain how this happens.
2. Do you think that water can break rocks? How?
3. What happens to water when it freezes? [gets hard, takes up more room]
4. What if water were frozen inside a rock? Explain how you think the frozen water could break the rock?
5. Look at our prediction graph. What does it tell us?

6. Look at our model rocks and the brick. What happened to the water inside? Why did they break? [the water got hard (froze), took up more room, and broke them]
7. Explain how you think this could happen in nature with real rocks.
8. When rocks are broken by water freezing, we call this *weathering*. There are other ways that rocks can be changed. Can you think of other ways? [water running over rocks, water dripping on rocks, earthquakes moving rocks]
9. Look at our actual results graph. How many of our model rocks were broken by the frozen water? How many were not broken?
10. Look again at the model rocks that were not broken. Why do you think these rocks did not break? Was there anything different about these models compared to the ones that broke? [less water inside, more dough around them, etc.]
11. If you did this activity again, what would you do differently?... the same? How could you make the activity better?
12. Is there anything you would like to know more about in this activity? Explain.

Extensions

1. Test different materials for making the model rocks. Are some materials stronger than others? Are some rocks harder than others and thus stronger than others?

Simulated sandstone* recipe

Large paper cup
Cementing solution (2 parts water to one part Epsom salt)
Sand

- Add some sand to the bottom of the large paper cup. Put in the water balloon which is filled with water to golf ball size. Pack the sand around the balloon with your hand, completely covering the balloon.
- Add cementing solution until all the sand is wet.
- Put the cup in a warm place until the sand dries completely.
- Carefully tear away the paper cup.
- Place the model rock into the freezer.
  *Note: Natural sandstone is made when small pieces of rock or sand are packed together in layers. Water containing minerals seeps in between the pieces and then evaporates. The minerals that are left behind cement the particles together into a larger rock.
2. To help reinforce the concept that water expands when it freezes, you may want to have students fill film canisters with water, snap on the lids, and put in the freezer along with the rocks.
3. Ask students if they have ever placed a soda in the can in the freezer. Have them describe what happened. Ask how the frozen soda applies to what happened to their model rocks.

Make a Model ROCK

You will need: small water balloon, plaster of paris, water, **or** salt/flour clay ½ cup salt + 1 cup flour, water

Do this:

1. Make a very small water balloon.

2. Cover the water balloon with plaster of Paris or salt/flour clay. Make it look like a rock.

   > This is like water seeping inside a rock.

3. Let your model rock harden. Observe. Compare it to a real rock.

   How is it like a rock?

   How is it not like a rock?

4. Put your model rock inside a freezer for 1 night.

5. What do you think will happen?

6. What happened?

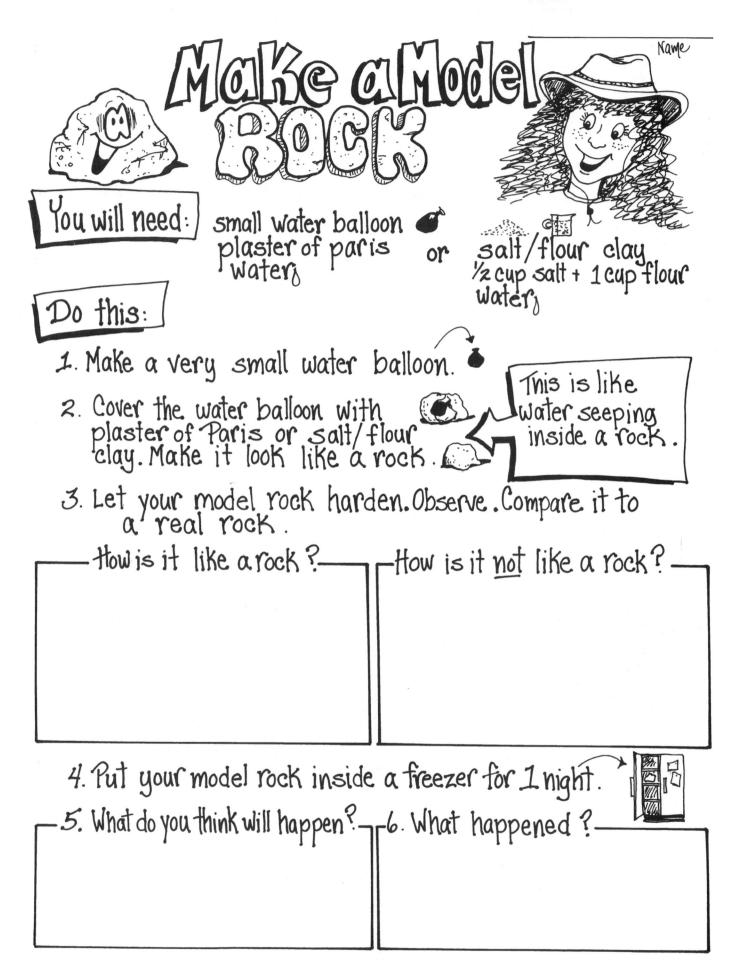

48

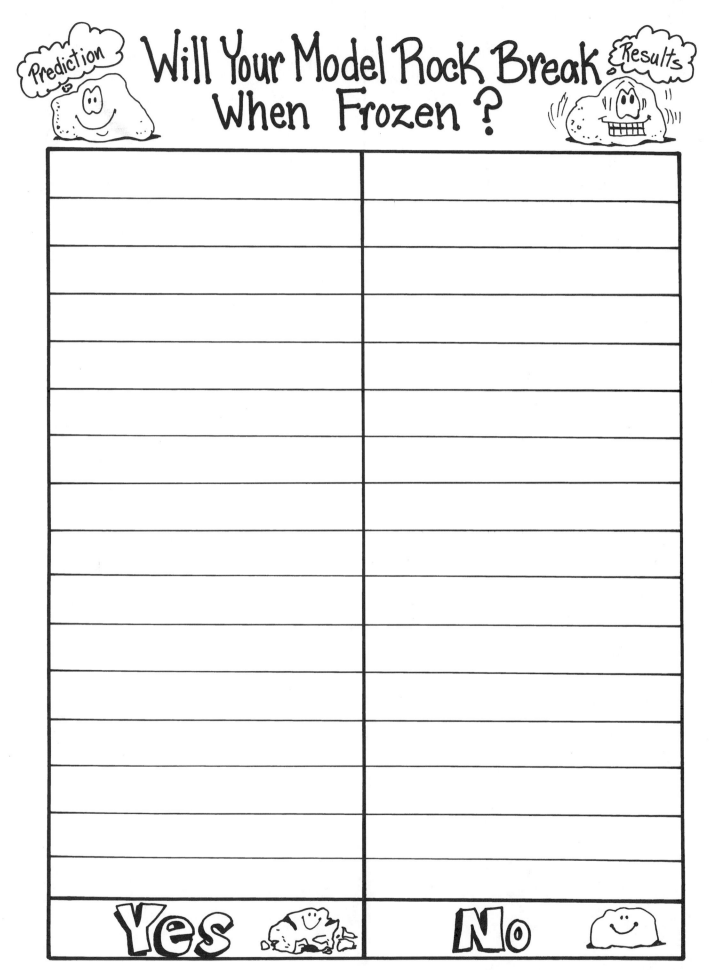

Will Your Model Rock Break When Frozen?

Prediction · Results

Yes	No

Prediction

I think the model rock will look like this after freezing.

Observe

The model rock looked like this before freezing.

Results

This is what the model looked like after freezing.

Open this up to see what I learned.

Ice Breakers

Scientist: _____

Topic
Erosion

Key Question
How do wind, water, and ice change Earth's surface?

Focus
Students will use models to observe that rocks are weathered into sands and soils.

Guiding Documents
NSE Standards
- *The surface of the earth changes. Some changes are due to slow processes, such as erosion and weathering, and some changes are due to rapid process, such as landslides, volcanic eruptions, and earthquakes.*

Project 2061 Benchmarks
- *Change is something that happens to many things.*
- *Waves, wind, water, and ice shape and reshape Earth's land surface by eroding rock and soil in some areas and depositing them in other areas, sometimes in seasonal layers.*
- *In doing science, it is often helpful to work with a team and to share findings with others. All team members should reach their own individual conclusions, however, about what the findings mean.*

Science
Earth science
  erosion
  deposition

Integrated Processes
Observing
Collecting and recording data
Comparing and contrasting
Predicting
Applying

Materials
Fine sand or dirt
Shoe box lids
Rectangular pans (9" x 12")
Paper cup
Masking tape
Water
Drinking straws
Safety goggles
Fine grained colored sand

Optional (see *Management 5*):
  gravel
  clay
  ice chunks

Background Information
Weathering, erosion, and deposition are three of the processes that change the Earth's landscape. Mountains and hills are gradually flattened and valleys widened into huge plains through these processes.

Weathering is a term that includes all the changes in rock materials that result from their exposure to the atmosphere. It transforms solid bedrock into small fragments that can be removed by agents of erosion.

Erosion is the wearing down of Earth's surface by natural forces. The tools (agents) for the changing of Earth's crust can be wind, water, and ice. Whatever the method, Earth is constantly being altered. The process of erosion is often very slow and difficult to observe. But over thousands of years, the mountains and hills have been worn down and rivers have widened their valleys into broad plains.

Water is by far the most powerful agent of erosion. More than one-quarter of the annual precipitation falling onto the continents runs off into the ocean via rivers, streams, etc. Rivers carry away rocks and soils, eroding the mountains and hills and carving out valleys and canyons. Streams of water roll materials downhill onto the lowlands or out to sea. Such erosion is often easily seen in coastal areas and along rivers and streams where noticeable amounts of land can be lost each year. Water continuously erodes and changes Earth's surface.

Wind erosion occurs mostly along the ground surface. Wind carries off soil and small rocks. The surfaces of boulders and rocks may eventually be pitted or worn smooth by the flying dust and sand grains. During the 1930's, large amounts of topsoil were lost when areas of the Middle West received little rain. Plants died and could no longer anchor the soil when winds blew. This area and period of time became known as the Dust Bowl.

Snow collects in hollows of mountains. As the weight of the snow builds up, it becomes compressed and forms ice. As more snow is added and turns to ice, gravity and the weight of the ice causes it to slowly move downhill; thus a glacier is formed. Glaciers carry embedded rocks and soil a great distance before dropping them. The glaciers

also act like bulldozers by pushing rocks in front of them. Rocks stuck in the bottom and sides of glaciers scrape, scratch, and dig into the rocks beneath the glacier. Glaciers can also cause erosion as the ice melts and slowly flows downhill.

Deposition is the laying down of eroded materials. It is evident in the bends of rivers and the deltas that form as major rivers drain into the oceans. Materials deposited by winds are found in the lee of objects and structures. Glacial deposits result as the glaciers melt leaving ridge-like hills.

Although erosion is a natural process, people can increase the effect by clearing land of vegetation or improper cultivation. They can also slow down the process by planting cover crops on bare land, terracing land, building wind breaks, etc.

Management

1. This activity is divided into several parts which investigate different agents of erosion. It is suggested that these parts be done over an extended period of time.

2. Look around the school ground or neighborhood for places where wind, water, or people have eroded the area. For real-world examples of *Water Erosion*, if there are no gutters or drain spouts, simply pour a bucket of water over some loosely packed soil so that students can get an idea of what happens with that process.

3. For erosion simulations, use fine sand if possible; it moves more easily than coarser-grained sand.

4. During *Wind Erosion* simulations, a fan or a student blowing through a straw (use the small coffee stir sticks) can move the sand. Blow-dryers are very dramatic; however, caution must be used to protect students' eyes. Safety goggles are strongly encouraged!

5. The results of glacial erosion are not evident in many environments. Students will not be able to find evidence of such on most school grounds. As a result, the teaching and learning in *Part 4: Glacial Erosion* is left as an option for primary students. If you wish to teach it, make large chunks of ice the night before the lesson. Use half-pint milk cartons and add sand and bits of gravel before freezing.

6. Select some objects in the classroom that students can pick up, carry, and deposit elsewhere to show change. Some suggested items are chairs, books, plants, etc.

7. Throughout all parts of the activity, emphasize the processes of *pick up, carry,* and *deposit.*

Procedure

Part 1
Introduction

1. Introduce the term *erosion* by doing some class role playing. Tell the students that they are going to change the room a bit. Inform them that the student doing the change will be called *Agent Erosion*. Tell them that the job of Agent Erosion is to pick up, carry, and deposit (put things down) items from around the room. Inform them that for the first few times, you will be the director and will tell Agent Erosion what to pick up, where to carry it, and where to deposit it.

2. Choose a student to be Agent Erosion. Direct him/her to pick up, carry, and deposit some items from around the room. Have different students repeat this procedure. Emphasize the processes and what changes resulted in the room. Select a student to be director and continue the role play.

3. Take the students outside to look at a door mat. Ask them what they see on or under the mat. Ask them how the dirt (grass, sand, etc.) got there. Ask them if they think Agent Erosion has been there. [This is a case of pick up, carry, and deposit. The soil, etc. was picked up on their shoes, carried to the area, and deposited when they scraped their feet. They were the agents of erosion.]

4. Tell the students that in nature the processes of pick up, carry, and deposit change the way Earth looks. Ask them what they think does the picking up and carrying.

5. If it is possible, find areas on the school ground where erosion can be shown—gullies formed by running water or hard-packed paths formed by human feet, piles of dust in the corners of buildings.  Ask the students how many of them have experienced having dirt or sand blown in their faces. Ask them which agent of erosion is picking up and carrying that dirt and sand.

6. Back in the classroom, list the agents of erosion [wind, water, ice, people] on the board. Also list places where the students have seen erosion. [steep hillsides, seashore, river-bottom land, ice-covered land, bare ground where people have worn a path.]

7. Ask where the soil and rock go when they are eroded.

52

Part 2
Wind Erosion

1. Review what the students learned about erosion. Tell them that they are going to see how wind erodes sand and dirt.

2. Place a thin layer of sand or dirt in a shoe box lid (or a flat pan) and place the container where all the students can observe it.

3. Discuss what the students think will happen when the sand is blown by wind.

4. Have students wear safety goggles, if available; otherwise, caution them to be very careful not to get any sand and dirt particles in their eyes. Tell the students that you are Agent Wind and you will create a wind by blowing gently through a straw. Have them predict what will happen.

5. Ask them what actually happened when Agent Wind blew across the sand and dirt. Have them point out where the particles were deposited.

6. Suggest the students role play Agent Wind Erosion. Stress that they must blow gently or they might get sand in their eyes. Urge them not to inhale any sand or dirt through the straw. Give the students a tray with sand and dirt on it. Let them blow gently through straws to move the sand. Tell them to observe what happens.

7. After they have blown the sand and dirt, have them describe what happened and relate it to the wind blowing outside.

8. Ask the students what objects wind can move. [small rock particles, sand, light soil]

9. Place rocks on top of the sand. Ask the students what happens to the soil around and under the rocks when the wind blows.

10. If available and time permits, use other things such as a fan or blow-dryer to simulate the wind.

11. Invite students to go outdoors to look for evidence of wind erosion. (Be sure to include observations of litter.)

Part 3
Water Erosion

1. To introduce this part of the lesson, take the students outside to look for evidences of water erosion on the school grounds. Look for three types of terrain around the school ground: bare soil, an area covered with vegetation, and concrete or asphalt. Pour the same amount of water on each of the terrain types and see what happens. Have the students compare and contrast what happens in the three situations.

2. Tell the students that they are going back into the classroom to see how water erodes sand and dirt.

3. Place the soil and dirt mixture in the rectangular pan about 5 cm deep. If colored fine sand is available, spread a thin layer over the mixture. This will help the students see that the sand moves due to erosion.

4. Prop up one end of the pan about 10 cm. Poke a hole in the bottom of a paper cup. Cover the hole with masking tape. Fill the cup with water.

5. Ask the students what will happen when you hold the cup over the dirt pan and pull off the tape. [The water will pour out onto the dirt.] Ask what they think will happen to the dirt.

6. Invite a student to be Agent Water Erosion and pull off the tape. Ask students what this is like in nature. [rain]

7. Have them describe their observations and relate them to the water erosion they saw on the playground.

8. Allow time for many students to role play Agent Water Erosion.

Part 4
Ice and Glacier Erosion (optional)

1. Relate to the students that a glacier is a huge amount of ice and snow that has collected over many, many years. Tell them that when the ice gets thick enough, it starts moving down the side of the mountain where it was formed. The tremendous weight of the glacial ice erodes the land by gouging and plowing up the ground and moving the rocks around. Glaciers can move large amounts of soil and large rocks many miles.

2. If possible, show pictures of a glacier, or a picture of the mountains in Yosemite National Park to explain the erosion of glaciers.

3. Flatten a piece of clay.

4. Remove the carton or paper cup from the "glacier" that was frozen prior to this part of the activity.

5. Ask a student to be Agent Ice Erosion. Direct the student to move the chunk of ice over the clay.

6. Have the students describe what has happened to the clay. Discuss the large indentations and scratch marks left on the clay. This is similar to what happens in nature when a glacier moves over the ground.

7. Allow other students to role play Agent Ice Erosion and move the ice over the clay.

Discussion

1. In the first part of the activity on erosion, how did you change the room? (Stress the use of the words *pick up, carry,* and *deposit.*)

2. Describe how nature does the same thing to our Earth.

3. What erosion do you see as you walk around the school ground?

4. How does water change the Earth?

5. How does wind change the Earth?

6. If we think of our pan of dirt and sand as a mountain and we keep pouring water on it, what will happen to the mountain?

7. Can you see signs of erosion everywhere? Explain.

Extensions

1. Have the students create a miniature mountain with damp soil in a flat rectangular box (9" x 12"). It will represent a model of a mountain without any trees or vegetation on it. Ask what they think will happen when rain falls on the mountain. Invite the students to use the paper cup with a hole to pour "rain" on the mountain. Urge them to describe what happens to the mountain. Where does most of the erosion take place? Where does the eroded soil go? Discuss any changes that may have occurred on the mountain. Ask if they can identify miniature streams and canyons.

2. Repeat this erosion demonstration and add leaves, grass, twigs, and rocks to various parts of the mountain. Do these objects increase or decrease the erosion on the mountain?

3. After the soil is wet from the erosion by water, blow a strong wind over the wet landscape. How does the moist soil affect the wind erosion?

4. Ask students to brainstorm ways they could prevent wind and water erosion. Relate this to the real world.

Curriculum Correlation

Language Arts
Have the students write a story about a rock that starts in the high mountains and is gradually eroded by ice, water, and wind as it moves from the mountain top to the shores of the ocean.

Home Link

Have the students look outside their homes for evidences of erosion and explain to their parents what caused the erosion.

Wind Water Ice

Agent Erosion

How does erosion change land?

Pick one agent of erosion to test. Wind Water Ice

┌─ Before Erosion ──────────────────────────────┐
│   │
│   │
│   │
│   │
│   │
│   │
└───┘

┌─ After Erosion ───────────────────────────────┐
│   │
│   │
│   │
│   │
│   │
│   │
└───┘

55

Soil Study

Topic
Soil and its contents

Key Questions
1. What is in soil?
2. How are soils different?

Focus
Students will observe soil samples to discover that soil is made from small bits of rock and once living things.

Guiding Documents
NSE Standards
- *Earth materials are solid rocks and soils, water, and the gases of the atmosphere. The varied materials have different physical and chemical properties, which make them useful in different ways, for example, as building materials, as sources of fuel, or for growing the plants we use as food. Earth materials provide many of the resources that humans use.*
- *Soils have properties of color and texture, capacity to retain water, and ability to support the growth of many kinds of plants, including those in our food supply.*

Project 2061 Benchmarks
- *People can often learn about things around them by just observing those things carefully, but sometimes they can learn more by doing something to the things and noting what happens.*
- *Often you can find out about something big by studying just a small part of it.*

Science
Earth science
   soil

Integrated Processes
Observing
Collecting and recording data
Comparing and contrasting
Inferring
Graphing
Communicating

Materials
Soil (see *Management*)
Sand (from the sandbox)
Bits of leaves, grasses, and dead insects

Clear container
Water
Sorting tools: strainer, tweezers, toothpicks
Magnifying lenses
Plastic bags
Plastic trays
Newspapers
Pots for soil
Radish seeds
Small jars with lids
Large paper for brainstorming chart

Background Information
Soil is a combination of small rock fragments and organic material. It is formed when the underlying rock is weathered into small fragments and mixed with organic matter. Soil is the foundation of life on land. Land animals and humans depend on plants for food and land plants need soil to grow.

Soil makes up the outermost layer of our planet. Beneath the thin layer of soil is bedrock. As plant roots, lichen, freezing and thawing, and wind and rain break up the rock, soil is gradually being formed. It can take up to 500 years to form one inch of topsoil.

Topsoil is the most productive soil layer. It has varying amounts of organic matter, minerals, and nutrients. The organic material in the soil is composed of decayed animal and plant material. The amount of organic material present determines the richness of the soil. Without organic material, soil would be lifeless, broken-down rock. Many organisms that inhabit the soil spend their lives breaking down dead animals and plants. The inorganic materials are made up of particles of rocks, a mixture of sand, silt, and clay.

Much of Earth's surface is covered with broken and crumbly rock materials that include sand, gravel, silt, and clay. These different-sized particles give soil texture.

Management
1. It is best if students can participate in gathering soil samples from different parts of the school ground. When you take the students outside to collect the soil samples, get some from a barren play area and

some from a weedy patch. If there are no patches with vegetation, you may have to augment some soil samples since the soils on school grounds normally are very poor in humus. Another option for obtaining soil samples is to ask the students to bring some soil from their yards in plastic bags.

2. Sand from the sandbox on the school ground can be used when students make soil.

3. Living animals such as worms, earwigs, and sow bugs may be in the soil. Before these animals crawl away, have students draw pictures of them.

4. Reproduce the pages of *Soil Study*, cut them apart, and organize them as you wish. Staple the sheets together to make a recording book for the students.

Procedure:

Part 1
A first look at soil

1. Tell the students that the best way to understand soil is to look at it. Provide them with magnifying lenses.

2. Take the students outside and walk to various areas where the students can dig and examine the soil. Find areas that have bare ground and also an area where some vegetation is growing if possible. Ask them to look and describe what the ground under their feet looks like. Encourage them to observe any plants that are growing, rocks, dirt, ants (or other living creatures), twigs, and leaves.

3. Encourage them to feel and smell the soil.

4. When you return to the classroom, ask the students these questions and record on a chart. What is soil? What did you find in the soil? What do you think soil can be used for? Think of all the reasons we need soil and make a list of them.

Part 2
Observing soil samples

1. Provide each group of students with a plastic bag of soil and a plastic tray. Have them feel the soil. Ask questions such as: What do you observe? How does soil feel, smell, and look?

2. Encourage the groups to use the sorting tools to separate the different types of materials in the soil. Ask questions such as: What does the soil have in it? Have them make separate piles of roots, seeds, rock, etc. and ask how many different types of materials their samples contain.

3. Direct them to compare their sample with a neighboring group. Ask them how the samples are the same and different.

4. Give the students magnifying lenses to observe the soil and other objects they found in it.

5. Have the students record their observations by filling in the *Soil Study* booklet.

6. Let the students compare their soil displays with other groups' samples. How are they the same? How are they different?

7. Discuss how soil is used. Ask students to explain why soil is important. Tell them they can record their answers in the *Soil Study* booklet.

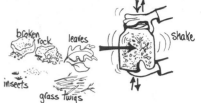

Part 3
Making soil

1. Have the students collect some sand from the school sandbox if one is available.

2. Put a thick layer of sand in the container.

3. Break up some leaves, grasses, stems, and twigs into tiny pieces. If necessary, grind this material between two rocks.

4. Add the plant material and dead insects (such as bees, ants, flies, wasps, and etc.) to the sand. Use about the same amount of this material as the sand.

5. Add water to the jar, put a lid on, and shake gently to mix the materials.

6. You have made soil! Ask the students if it looks like any soil they have seen before. Discuss how they could make the soil look more like the soil they observed. Have them smell their soil and compare/contrast the odor with the soil on the school yard.

7. Tell the students that they are going to set up a comparison of the soil they have made and the soil they got from the school yard to see if they can grow radishes in them. Inform students that they will put each type of soil in a pot, dampen the soil, and plant some radish seeds in each pot. Have them predict the growth of the seeds in each pot and explain any differences in their predictions.

Discussion

1. What is your definition of soil?

2. What makes one kind of soil different from another kind?

3. How did the soil feel?... smell?... look?

4. Describe the kinds of things you found in the soil?

5. What words would you use to describe the soil you collected on the school grounds? Is it like the soil in your yard?

6. Have you ever looked at the soil at home? Does it look different than the soil you used in this activity?

7. Explain how you think soil is formed.

8. When you made soil, how long did it take?

9. What did you learn about soil that you didn't know before?

10. What might happen if there were no leaves or grass in the soil?

11. Why is it important for us to be careful of our soil?

Extensions

1. To show how rocks change, fill a one-pound coffee can about one-quarter full of small rocks. Put the lid on the can. Have the students pass the

can around and ask each student to shake the can vigorously. After each student has had a turn, open the can and pour the rocks out. Observe the rocks and discuss the changes. Why is shaking the rocks like the action of water and wind on rocks? [both actions wear down rocks]

2. Discuss with your class the benefits of composting. Begin a composting pile by gathering leaves. Add water and watch them change.

3. Fill two small jars two-thirds full of water. Add a small amount of the student-made soil to one jar and a small amount of the school yard soil to the other. Shake the jars and watch what happens to the materials. Let the jars sit until the next day. Observe and compare.

Home Link

Have the students bring soil samples from their own yards. Have them look at the samples with magnifying lenses to find any differences in these samples from those found on the school yard.

Soil Study

Geologist: _____
Dated: _____

I Saw

I smelled

I felt

Soil looks like this
when magnified.

This is where I found
my soil.

Soil is used for

Soil is important
because

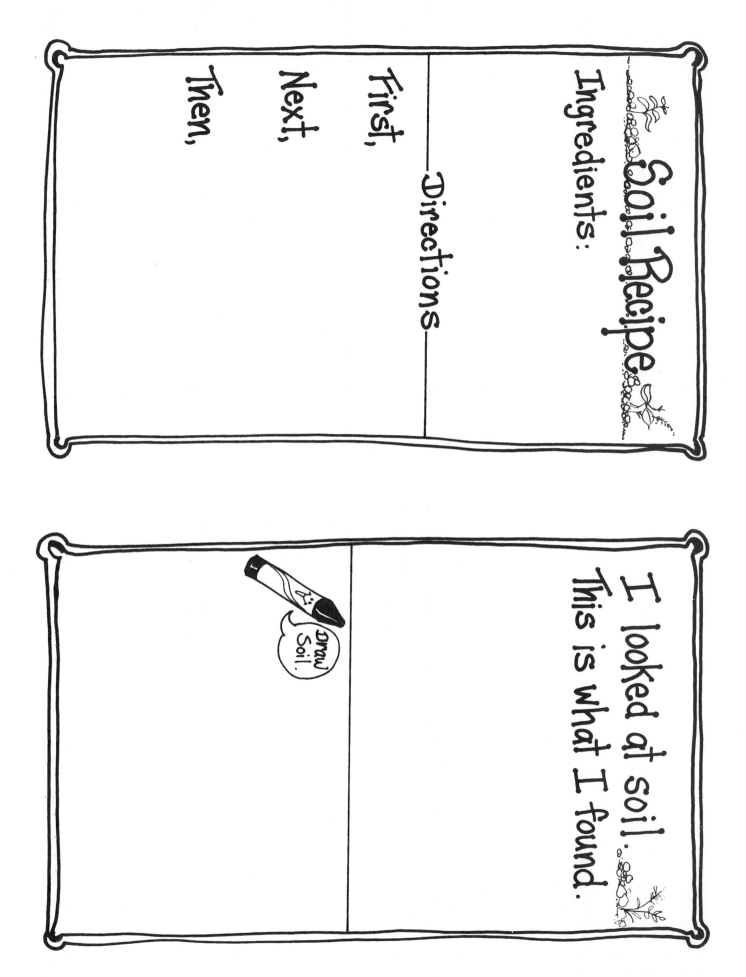

Soil Recipe

Ingredients:

Directions

First,

Next,

Then,

I looked at soil.
This is what I found.

Draw Soil.

Here is a sample of
the soil I studied.

glue

These are the colors I see.

Compare

The soil I found has

The soil I made has

62

Sandpile

Topic
Sand

Key Question
How would you describe sand?

Focus
Students will observe that sand is made from small bits of different rocks.

Guiding Documents
NCTM Standards
- *Describe, model, draw, and classify shapes*

NSE Standards
- *Earth materials are solid rocks and soils, water, and the gases of the atmosphere. The varied materials have different physical and chemical properties, which make them useful in different ways, for example, as building materials, as sources of fuel, or for growing the plants we use as food. Earth materials provide many of the resources that humans use.*
- *Employ simple equipment and tools to gather data and extend the senses.*

Project 2061 Benchmarks
- *Objects can be described in terms of the materials they are made of (clay, cloth, paper, etc.) and their physical properties (color, size, shape, weight, texture, flexibility, etc.).*
- *Magnifiers help people see things they could not see without them.*
- *Describing things as accurately as possible is important in science because it enables people to compare their observations with those of others.*
- *Chunks of rocks come in many sizes and shapes, from boulders to grains of sand and even smaller.*

Math
Observing geometric shapes

Science
Earth science
  sand

Integrated Processes
Observing
Classifying
Comparing and contrasting
Recording data
Communicating

Materials
White glue
Samples of sand
Magnifying lenses
Containers for sand
Microscope and student prepared slides (optional)

Background Information
Sand is but the Earth's crust in miniature form. Earth's rocks eventually are weathered and become sand. This all is part of the rock cycle. The actions of oceans, river, glaciers, rain, and wind are constantly eroding Earth's rocks into smaller and smaller pieces. First the rocks are eroded into boulders, then the boulders are gradually broken down into pebbles and gravel. The grinding of these objects against each other makes smaller and smaller particles which result in sand.

Sand occurs on ocean and lake beaches, in deserts, and in riverbeds. When observed from a distance, sand may appear to be a solid color, but when carefully observed with a magnifying lens, particles of several different colors are seen. One common mineral in sand is quartz. Pure quartz crystals have a glassy luster and are colorless, but slight impurities can produce a variety of colors.

Sand is very important to us. It is used for building, paving, and making concrete. Sand is used to make pottery and sandpaper. Sand is also used in making glass.

Management
1. Gather the sand beforehand. A good source of sand for this activity is the school sandbox, if you have one. Other sources of sand are: beaches, streambeds, hardware stores, pet stores, or building supply stores.
2. To facilitate observations, have students glue the sand to the paper.

Procedure
1. Gather the sand and place it in containers. Pass the containers around and let the students feel the sand. Have them observe the colors of the particles and the sizes of the grains.
2. Distribute the activity pages. (If more than one sample of sand is used, distribute another copy of the activity sheet so comparisons of the two samples can be made.)
3. Ask the students to spread a thin layer of white glue in the circle, sprinkle a layer of sand on top of the glue, and let it dry thoroughly.

4. When the glue is dry, invite the students to feel the sand and describe how it feels.

5. Distribute hand lenses and encourage students to use them to observe their sand.

6. Ask students to describe the shapes and colors of sand grains they see. Describe the mineral quartz to them by telling them that it appears glassy and is often colorless. Ask them if they see any quartz in their samples.

7. Have students put some sand into the film canisters, filling them about one-third full. Tell them to shake the canisters and describe the sounds. If you have varieties of coarse and fine sand available, ask the students to compare and contrast what they hear and see.

8. If sand is available from various areas, divide the students into groups and have them observe, compare, and contrast the characteristics of two samples. You may even want to set up a mystery sample of sand and ask students to match it using the set of samples you have available.

9. Optional: Have students prepare microscope slides by cutting 1" x 3" strips of tagboard. Fold these strips in half (1" x 1.5"). Along the fold line, cut out a triangular shaped piece. Open the strip to see a diamond-shaped cutout area. Cover this area with a piece of transparent tape. Sprinkle a few grains of sand onto the sticky side of the tape. This now becomes the microscope slide.

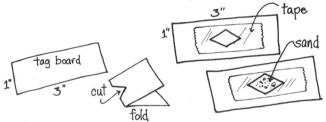

10. Have the students view the slide under the microscope looking for various geometric shapes in the grains of sand. You may wish to have students draw three or four grains of sand on the reverse side of their handouts.

11. Discuss what sand is. [grains of worn down or disintegrated rocks] Talk about the uses of sand. [for sand paper, sandstone for making buildings, making glass, etc.] If appropriate, invite students to research the use of sand.

12. End with a whole class discussion, sharing what everyone has learned about sand in this activity.

Discussion

1. How does the sand feel? Does all sand feel the same? Explain.

2. Describe what sand looks like under the magnifying lens.

3. What colors do you see? Do you see one color more than any other?

4. Describe sand using your sense of sight, hearing, and feeling.

5. Explain what you know about how sand was formed.

6. What did you learn about sand today?

Extensions

1. Provide the students with magnifying lenses to observe and compare sand, salt, and raw sugar.

2. Have the students compare their sample of sand with a piece of sandpaper.

3. Gather and compare sand from various locations in the country or world.

Curriculum Correlation

Art

1. Draw a design by drizzling white glue over a piece of colored paper. Sprinkle sand over it. Let the sand painting dry.

2. Have the students draw a picture of the desert and cover the sandy areas with glue. Sprinkle sand over the glue.

3. Make your own sand sculpture by using 500 ml sand, 250 ml cornstarch, and 250 ml water. Mix together and heat. When the mixture is thickened, remove from the heat and let cool. Mold your own sculpture. It will harden as it dries.

4. Make different colors of sand by mixing clean dry sand and several drops of food coloring together. When dried this sand can be put into wide-mouthed jars in layers, or made into interesting designs.

Social Studies

1. Ask the students to research the sand paintings done by the Native Americans of the Southwestern part of the United States.

Sandpile

Use a hand lens to observe sand.

Spread glue in the circle.

Sprinkle your sand on top.

Let it dry.

1. The sand feels _____.

2. The sand looks _____
   _____.

3. The colors of the sand are _____
   _____.

4. The sand is made of _____
   _____.

65

The Earth Has What We Need!

The earth has what we need.

Topic
Earth's resources

Key Questions
1. What rock and mineral materials do we use that come from the Earth?
2. How can we reduce our use of Earth's materials?

Focus
The students will identify rock and mineral materials in the classroom and around the school that have come from the Earth.

Guiding Documents
NSE Standards
- *Earth materials are solid rocks and soils, water, and the gases of the atmosphere. The varied materials have different physical and chemical properties, which make them useful in different way, for example, as building materials, as sources of fuel, or for growing the plants we use as food. Earth materials provide many of the resources that humans use.*
- *Resources are things we get from the living and nonliving environment to meet the needs and wants of a population.*
- *Some resources are basic materials, such as air, water, and soil; some are produced from basic resources, such as food, fuel, and building materials; and some resources are nonmaterial, such as quiet places, beauty, security, and safety.*
- *The supply of many resources is limited. If used, resources can be extended through recycling and decreased use.*

Project 2061 Benchmarks
- *Objects can be described in terms of the materials they are made of (clay, cloth, paper, etc.) and their physical properties (color, size, shape, weight, texture, flexibility, etc.).*
- *Many materials can be recycled and used again, sometimes in different forms.*
- *People can use objects and ways of doing things to solve problems.*
- *Magnets can be used to make some things move without being touched.*
- *Some materials can be used over again.*

Science
Earth science
conservation of natural resources

Integrated Processes
Observing
Comparing and contrasting
Classifying
Collecting and recording data
Communicating
Inferring

Materials
Objects in the classroom
    made from rocks and minerals (chalk, doorknobs, paper clips, window glass, handles, etc.)
Aluminum can
Tin can
Plastic bottle
Magnifying lenses
Magnets *(optional)*

Background Information
Earth provides everything people need. It gives us food and the materials needed to make clothing and shelter. The fuels that furnish energy come from Earth's crust. Minerals for industry and buildings also come from the Earth.

There are both renewable and non-renewable natural resources. Some resources, such as forests, wildlife, and water, are renewable. With some exceptions, these things are considered replaceable in a relatively short period of time. Other resources, such as fossil fuels, are considered non-renewable because they cannot be replaced or their replacement is extremely slow.

All rocks are made of one or more minerals. Most minerals are a non-renewable resource because their formation takes millions of years. Metals are made from minerals found in rock. Humans use minerals to manufacture almost everything–paint, computers, glass, batteries, pencils, and televisions. We use materials from rocks to build homes, sidewalks, schools, and roads. Our bodies must have certain minerals in order to carry on life's processes. Even the food we eat grows in soil made from rock.

Management
1. Before teaching this lesson, take a walk around the school. Locate various examples of different rock and building materials (concrete, granite, gravel, sand, tile, asphalt, etc.) that you would like your students to see.
2. Look around the classroom to locate and identify for yourself various items and materials that are made of metal or rock materials.

3. Collect a box of recyclable trash such as: two-liter plastic bottle, tennis shoe, aluminum foil, aluminum can, paper, milk carton, plastic milk container.

Procedure

1. Take the class on a *Rock-Use Walk* around the school to observe how rocks and rock materials are used. Have the students point out rocks used in construction of buildings, rock walls, gravel on paths, concrete in the buildings and sidewalks, and sand in the sandbox. Emphasize that all these are rocks and rock products that are important in our everyday life.

2. Give each student or pair of students a hand lens and encourage them to examine closely the various rock materials. Discuss what they notice and help them identify the rock materials.

3. Back in the classroom, discuss with the students that the Earth is the only place we can get the materials that they saw and many other materials that we use in our everyday life.

4. Pass around an aluminum can so the students can handle it. Tell them that the can is made from a mineral called aluminum that is mined from the Earth, taken to a factory, and made into a silver gray material which is then used for this soda can and many other items. Ask them if they find anything else in the classroom that is made from aluminum.

5. Now show them a tin can. Explain that it contains a material called iron. Like the aluminum, iron is processed in a factory and made into many different objects. Inform them that iron is one of the minerals that has magnetic properties. Ask how they could find things in the room that contain iron. [Use a magnet.] Challenge them to find other objects in the classroom that are made of iron.

6. Encourage the students to assemble and illustrate their own *My Earth's Resources Book*.

Discussion

1. What rock materials do you see as you walk around the school?

2. As you look closely at the different rock material, what do you notice? How are they alike?... different? Are they all made of the same rock? How can you tell?

3. How are the rock materials you saw used in buildings? Why do you think they were used for that particular purpose?

4. What do you see in the classroom that you think is made of aluminum? How can you tell? What can you think of at home that might be made of aluminum? Why do you think aluminum might be a good material for making things?

5. What do you see in the classroom that you think is made of iron? How can you tell? [Use a magnet.] What can you think of at home that might be made of iron?

6. What do you see in the classroom that you think is made of plastic? How can you tell? What are some of the characteristics of plastic? What can you think of at home that might be made of plastic? Why might plastic be a good material for making things?

7. Think about jewelry. What different kinds of rock materials and minerals are used to make jewelry?

8. Did you know that you even eat a rock? Salt is a mineral that comes from the Earth. What other ways can you think of that we use rocks and rock materials every day?

9. Why do you think it is important to protect the resources that come from the Earth? What can we do here in class to help? What can your family do? What can YOU do?

Extensions

1. Discuss how some things that we throw away can be used again by recycling them at home or elsewhere. Show the students a box of recyclable trash. Tell them to pick one piece of trash and find a way to reuse it. Have them record what they think they can do with the trash on the recording sheet *Use it Again*.

2. Have the students take a magnet around the classroom and find objects that the magnet attracts. Record these objects on the sheet *Metal Search*. On the same sheet, record other objects in the classroom that are made of metal but are not attracted by the magnet.

3. Discuss care of our resources — Reduce, Reuse, Recycle. Begin a class recycling program. Have the students make a list of how they can reduce their use of materials such as paper and pencils.

4. Take a field trip to a landscape or building supply center to look at the variety of building materials that come from the Earth.

5. Show the students a piece of paper. Discuss that it comes from trees. The trees need soil from the Earth to grow. When the tree is grown, it is cut down and taken to the mill to make pulp which is eventually turned into paper.

6. Tell the students to look at their pencil. Every part of it comes from the Earth. Ask them if they can figure out where each part originally came from. [lead-graphite (carbon); wood-trees that grew in the soil; eraser-rubber plant; metal band-iron ore; paint-oil] Ask them if they can identify something else around their desk or in their that is made of a combination of materials from the Earth.

Curriculum Correlation

Language Arts

Have the students read *Readers' Theater*. Divide the class into six groups. Each group performs one section. Everyone chants the refrain which is indicated with a •.

Home Link

Have the students find out what materials from the Earth were used to build their homes. Make a class list or graph to show what they found.

My Earth's Resources

by:

We use resources that come from the Earth.
1.

Soil is used to grow food and it comes from the Earth.
2.

Copper is used to make pennies and it comes from the Earth.
3.

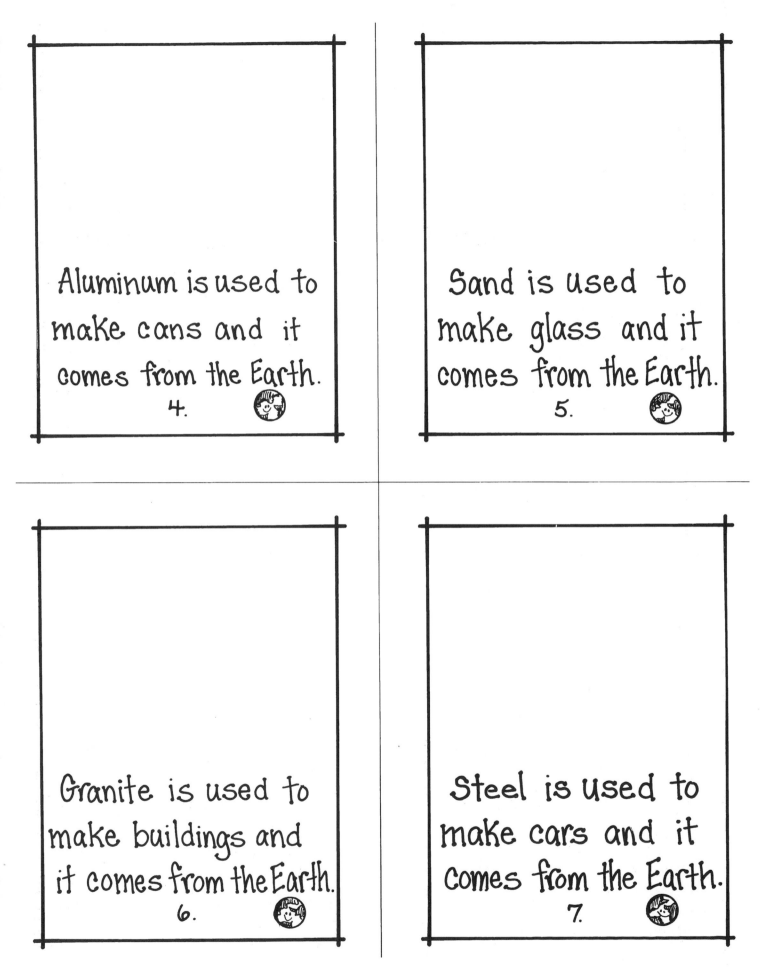

Aluminum is used to make cans and it comes from the Earth.
4.

Sand is used to make glass and it comes from the Earth.
5.

Granite is used to make buildings and it comes from the Earth.
6.

Steel is used to make cars and it comes from the Earth.
7.

Silver is used to make silverware and it comes from the Earth.
8.

Clay is used to make plates and it comes from the Earth.
9.

Gold is used to make jewelry and it comes from the Earth.
10.

We must take care of the resources that come from the Earth.
11.

Use it Again

Pick one piece of trash.
How can you use it again?

We make a lot of trash!

1. My trash is _____.

2. My trash looks like this.

3. I could reuse my trash like this.

Metal Search

Name _____

The earth has what we need.

Find metals we use.
Test them with a magnet.
Draw what you find.

Metal with iron	Other Metals

Readers' Theater
(a chant)

We Use Rocks and Minerals

•Rocks and minerals, rocks and minerals,
•How do we use them? Rocks and minerals?
Aluminum we use to make pie plates,
Cooking foil, and even soda cans.

•Rocks and minerals, rocks and minerals,
•How do we use them? Rocks and minerals?
Soil we use to grow our food, green grass fields,
Shady trees, beautiful plants, and flowers.

•Rocks and minerals, rocks and minerals,
•How do we use them? Rocks and minerals?
Clay we use to make our dishes, cups and
Saucers, bricks and sculptures, flower pots.

•Rocks and minerals, rocks and minerals,
•How do we use them? Rocks and minerals?
Gold and silver we use for jewelry:
Rings and necklaces, braces and bracelets.

•Rocks and minerals, rocks and minerals,
•How do we use them? Rocks and minerals?
Steel we use to make our nails, nuts and bolts,
Cars and ships, buildings frames, forks, knives, and spoons.

•Rocks and minerals, rocks and minerals,
•How do we use them? Rocks and minerals?
Sand is melted and made into glass for
Doors and windows, eyeglasses, telescopes.

•Rocks and minerals, rocks and minerals,
•That's how we use them, rocks and minerals!

—Sheryl Mercier

Further Studies of the Earth

The overall goal for this publication is that students gain an appreciation for the planet on which they live by developing an understanding of some of its various components. While young learners will not be able to conceptualize the layers of the Earth, earthquakes, and volcanoes, the following three activities are included for teachers who would like to introduce them at an awareness level.

What's Inside?
Quaking Earth
Volcanoes

What's Inside?

Topic
Earth's layers

Key Question
What does the inside of the Earth look like?

Focus
The students will, through the use of a model, discover what Earth's interior is like.

Guiding Document
Project 2061 Benchmarks
- *Tools are used to do things better or more easily and to do some things that could not otherwise be done at all. In technology, tools are used to observe, measure, and make things.*
- *A model of something is different from the real thing but can be used to learn something about the real thing.*
- *One way to describe something is to say how it is like something else.*

Science
Earth science
    Earth's interior

Integrated Processes
Observing
Classifying
Comparing and contrasting
Inferring
Collecting and recording data

Materials
Apples
Knife
Globe (physical relief)

Background Information
People have always wondered what is inside the Earth. Geologists cannot just open up the planet to see what is inside, nor can they X-ray it as doctors do to see the inside of a person's body. Scientists take measurements of events that occur on and in the Earth; they then attempt to explain them. For example, geologists use instruments to find out what happens as earthquake waves move through the Earth. These waves pass through different materials at different rates. Think about how difficult it is to walk in water in a swimming pool

compared with walking in air. This can be compared with some materials in the Earth's interior that slow earthquake waves or cause them to change direction. By analyzing earthquake waves, geologists have theorized that the interior of the Earth consists of layers.

They believe that there are three main layers: the crust, the mantle, and the core. *Crust* describes the relatively thin layer on the surface of the Earth that covers it like a skin. We know the most about this layer because we can make direct observations. Much of the crust is made of rocks and soil. In some places ice and snow cover the rocks. In  other places ocean water may cover the rocks. The second layer is called the *mantle*. It is believed to be thick and made of extremely hot rock, so hot that it often flows around like molasses. The third layer is the *core*, or the center of the Earth. It is thought to be liquid outside and solid inside. It is probably made up of a mixture of iron and nickel.

Young students are very curious about their world. Although they will not be able to conceptualize the layers of the Earth, they may be ready for an awareness level activity (or series of activities, if available) that helps them understand that there are many things they cannot directly observe. They can learn about some of these things from hearing what scientists believe and by looking at models. At this level, young students should not be assessed on layers of the Earth. They should be encouraged to enjoy the pursuit of the mystery.

Management
1. Have several apples ready for the class.
2. You may wish to divide this lesson into two parts. Discuss the globe and its features on the first day and the interior of the Earth on the second day.

Procedure
Part 1
1. Tell the students they will be geologists for this lesson. A geologist is a scientist that studies the Earth.
2. Show the students a physical relief globe. If they are familiar with the globe as a model, ask them to point out where they live; otherwise, explain to them that this globe is a model of the Earth. Tell them that different colors are used for different purposes: the blue stands for water, the brown (and/or green) for land, white for ice.

3. Have students identify different features.

4. Have students describe the shape of the model of Earth. Ask them why they think the globe is round. [because the Earth is round] Ask them if the Earth seems round when they look out the window. If appropriate, point out that Earth is so large and we live on such a small sec-  tion of it that we can't see all of it, so it looks flat to us. Explain further that pictures of the Earth from outer space show that it is ball shaped. Give the students the recording sheet *Our Earth* to record their observations of the globe.

Part 2

5. Ask the students what they think is inside the Earth. Let them share their ideas. You may even want to take them outside to stand on the ground. Invite them to imagine what is under their feet.

6. Using the recording sheet *What is inside the Earth?*, have them draw a picture of themselves standing on the surface of the Earth and illustrate what they think is under their feet.

7. Tell the students that scientists have spent many years wondering what is inside the Earth. Discuss with the students that the geologists can't peek inside the Earth nor can they dig a hole deep enough in the Earth to explore what is down deep inside. Relate to them that they often have to use instruments to measure things such as earthquakes to understand about the inside of the Earth.

8. Point out to the students that we use a globe as a model to look at the outside of the Earth, and that in this activity they will use an apple as a model of the inside of the Earth. Hold up the apple and ask the students what is inside an apple. [a solid area] Ask them what is on the outside of the apple. [a skin or peel] Ask if there any other layers inside the solid area of the apple? [the core] Cut the apple in half vertically and look at the three layers. Cut another apple in half, this time horizontally. Look at the apple carefully.

9. Compare the apple to the Earth. [They are both roundish and solid and have layers.] Tell the students that they are going to examine the apple even more carefully to see how it is like the Earth. Inform them that an apple and the Earth both have a thin skin or crust. The crust of the Earth is like the skin of the apple, it is very thin in comparison to the rest of the Earth. Inform them that this is where they live and where the oceans and mountains are and that underneath the soil and water is rock. Explain that even though the crust is thin, on our huge world it is still thick enough that scientists cannot drill through it to the next layer.

10. Ask students what is directly under the skin or peeling of the apple. [the flesh] Point out that the flesh of the apple is like the middle, or mantle of the Earth; both are made up of thick material, the mantle of the Earth is made of hot melted rock. Explain that this hot melted rock is so deep that we usually do not worry about it; however, sometimes it makes its way from the mantle to the surface of the Earth when it is forced out of volcanoes as lava.

11. Again ask students what is next in the apple as you move inward. [the core] Inform them that the Earth also has a core and that scientists think the very core of the Earth is solid rock surrounded by an outer core which is a thick, hot, liquid rock.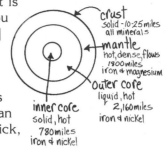

12. Have the students compare the layers of the apple with the layers of the Earth by recording the different layers of the apple on the recording sheet *Inside the Earth* and drawing a line between the layer of the apple with the layer on the Earth that it represents.

Discussion

1. If you were a geologist, what would you be doing?
2. Why is it hard to find out what is inside the Earth?
3. The Earth has several layers, which layer do we know the most about? Why?
4. What is the thinnest layer of the Earth?
5. Explain ways in which the layers of an apple are like the layers of the Earth.
6. Explain ways in which the apple is different than the Earth.

Extension

Make paper mâché globes as a class project. Blow up round balloons. Wrap them with strips of news-paper dipped in liquid starch. Add one more layer of news-paper strips dipped in starch and let dry. The next day, add another layer of strips dipped in starch and let dry again. When thor- 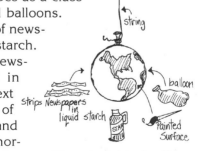 oughly dry, poke the balloons with a straight pin. Have the students draw and paint the continents and oceans on the sphere.

Our Earth

Geologist: _____

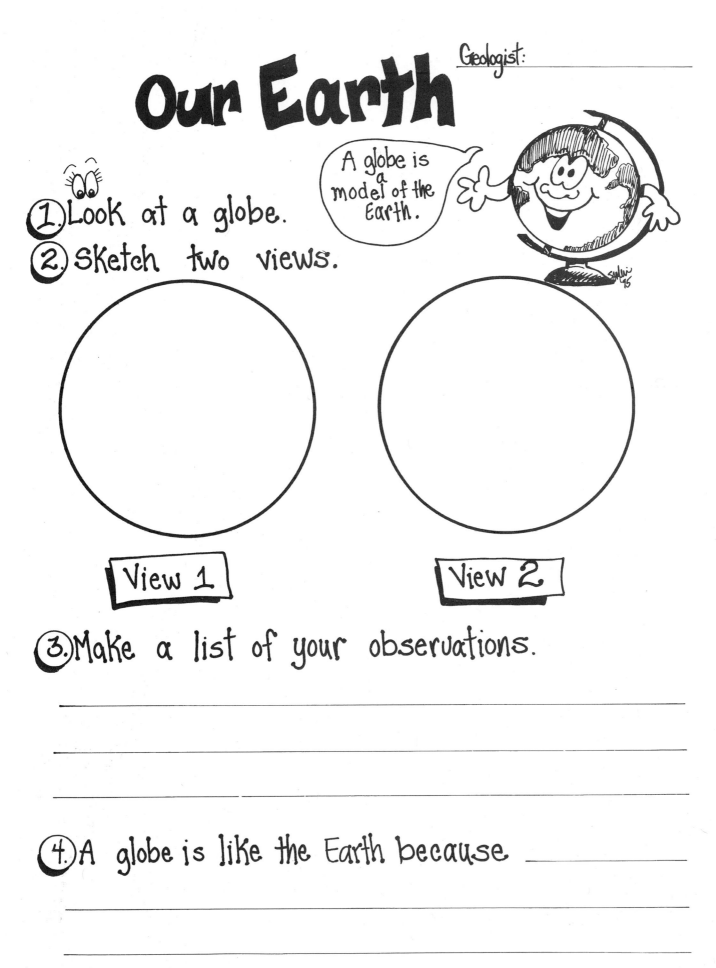

A globe is a model of the Earth.

1. Look at a globe.
2. Sketch two views.

View 1

View 2

3. Make a list of your observations.

_____

_____

_____

4. A globe is like the Earth because _____

_____

_____

What is inside the Earth?

Draw yourself standing on the ground.

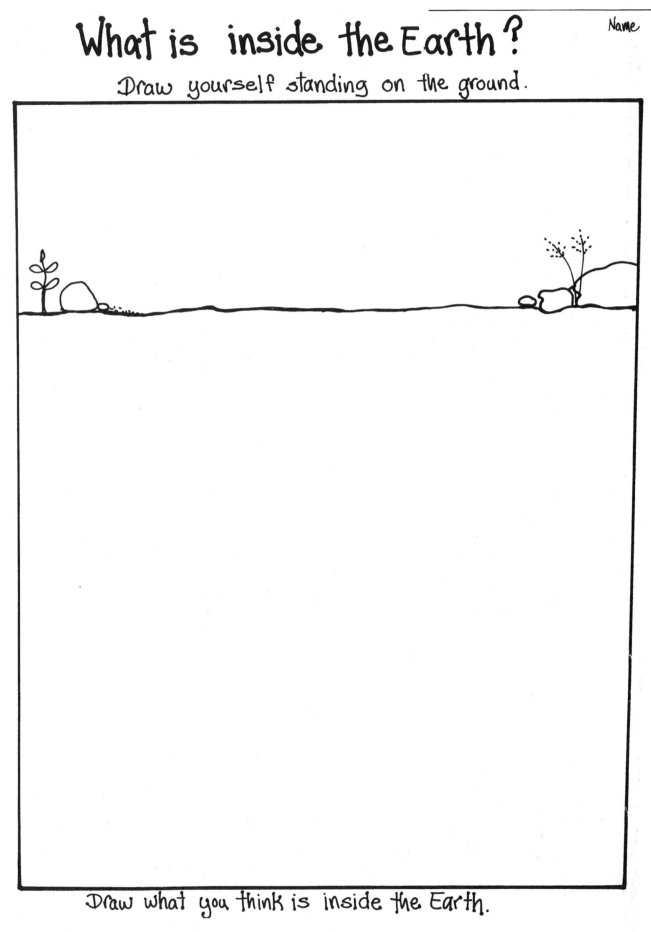

Draw what you think is inside the Earth.

Name _____

Inside the Earth

How is an apple like the Earth?

Geologist: _____

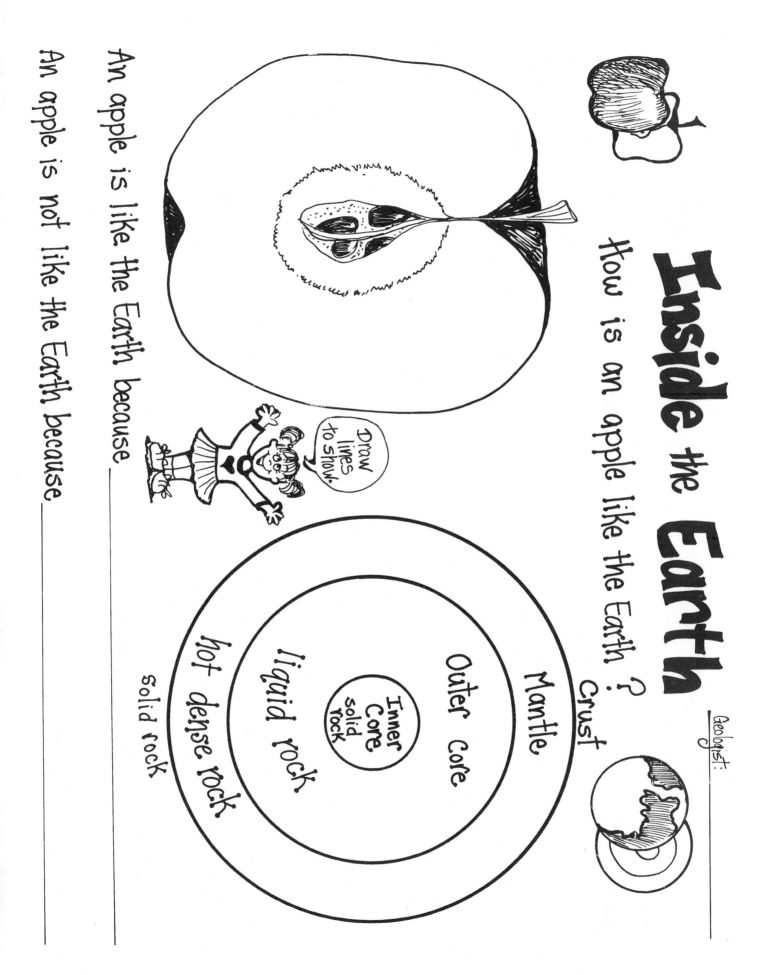

Draw lines to show.

Crust
Mantle
Outer Core — liquid rock
Inner Core solid rock
hot dense rock
solid rock

An apple is like the Earth because _____

An apple is not like the Earth because _____

79

Quaking Earth

Topic
Earthquakes

Key Question
What happens to Earth and structures on Earth when it quakes?

Focus
The students will do two investigations that will give them an idea what happens to the surface of Earth and buildings on Earth when it quakes.

Guiding Documents
NCTM Standards
- *Collect, organize, and describe data*

NSE Standards
- *The surface of the earth changes. Some changes are due to slow processes, such as erosion and weathering, and some changes are due to rapid process, such as landslides, volcanic eruptions, and earthquakes.*

Project 2061 Benchmarks
- *Some changes are so slow or so fast that they are hard to see.*
- *Change is something that happens to many things.*

Science
Earth science
  earthquakes

Integrated Processes
Observing
Collecting and recording data
Predicting
Classifying
Applying
Communicating

Materials
Part 1
For each group:
  a shallow cardboard box
  sand
  small blocks, paper cups (things that can be stacked)
For the teacher:
  a small desk or table that moves easily

Part 2
For the teacher:
  a pan of prepared gelatin dessert (see *Management*)
  small blocks, Unifix Cubes, or sugar cubes
  clear plastic wrap
  a small plastic glass
  water
  straws
  thread

Background Information
An earthquake is the trembling or shaking that results from the sudden release of energy in the rocks of Earth's crust. While there are many areas that scientists have indicated are more likely to have earthquakes, they are a geological phenomena that can occur anywhere.

It is believed that Earth's crust has seven large plates and about as many smaller ones that continually move. As the plates move, their edges rub against each other. As they move, the rocks that make up the plates bend, stretch, or get squeezed. From time to time, the edges of the plates may lock together. As the energy builds up between the two plates, it may be suddenly released, causing the rocks to break and shift. When the rocks move, they release a tremendous amount of energy in a short time. This sudden movement along a fault, a surface along a rock mass which has broken and moved, causes vibrations in the form of waves. The waves are often felt on the surface as an earthquake.

The waves created by the movement of the Earth can make the rocks vibrate back and forth, up and down, and from side to side sending the energy waves out in all directions from the center of the quake. These waves can cause damage to homes and businesses. The waves from some earthquakes are so small that they are just barely felt or not felt at all, but other ones are so strong that they can cause large buildings to crumble. Small or large, an earthquake is the result of some release of energy in the rocks of the Earth.

Management
1. *Part 1* of this activity has students use their hands to experience the release of energy as their hands are pressed together and slide past each other.
2. Prepare an earthquake model for *Part 1* by filling a shallow box or pan with dry soil or sand. Have an assortment of blocks, small milk cartons, and paper cups that can be stacked to represent homes and buildings.

3. For *Part 2*, prepare a pan of *Earthquake Gelatin* in advance.

```
        Earthquake Gelatin
2   6-oz. boxes of gelatin
2   one-serving envelopes of unflavored gelatin
4   cups boiling water
4   cups cold water
```

Empty the gelatin dessert and the unflavored gelatin into a 9x12 baking pan. Add the boiling water and stir until all the powder is dissolved, then add the cold water and stir to mix. Chill in the refrigerator for at least three hours or until set. Cover the top of the gelatin with plastic wrap.

4. *Part 3* emphasizes earthquake safety.

Procedure
Part 1

1. Introduce the topic of earthquakes by discussing with the students that the term *quake* means to tremble or shake.
2. Ask the students to tell the class what they already know about earthquakes. (Clear up any misconceptions at this time.)
3. Question the students if anyone has experienced an earthquake. Have those that raise their hands tell what happened to them and what they saw and felt.
4. Tell the students to hold their hands up, with the palms facing their face. Direct them to press the sides of their hands together as hard as they can like two large areas of Earth called plates pushing against one another. Then, still pressing very hard, ask them to try to slide one hand up along the other hand. Discuss whether moving their hands was hard to do and what happened when one hand finally slid. Ask if they felt a burst of energy. Relate this to a giant Earth plate that breaks free, and the burst of energy causes an earthquake.
5. Explain to the students that they will use a model to see what happens to the surface of the Earth and buildings on the Earth during an earthquake.
6. Fill a flat box or pan with sand and dirt.
7. Have the students pile the blocks, milk cartons, paper cups, and other objects on the box of sand and soil to represent a building.
8. Place the box on a table or desk.
9. Shake the table very gently, so that nothing happens to the structures.
10. Shake the table three more times, each time increasing the amount of force. The final shake should destroy the structures completely.

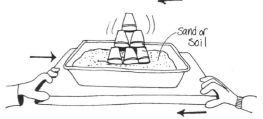

11. Compare the amount of damage that occurs as the shaking force increases. Invite the students to record what the structures looked like before the shaking started by drawing on the top half of the recording sheet *Earthquake Model*. On the bottom half of the sheet, draw what has happened to the structures after they have been shaken.

Part 2
12. Point out to the students that they are going to use a pan of *Earthquake Gelatin* as a model to see what happens during an earthquake. (see *Management*)
13. Hold the pan of gelatin firmly in one hand and gently tap the side of the pan with the other hand. Ask the students to describe what they observe. After some discussion and further demonstration, suggest to them that *waves* travel through the gelatin similar to the waves created by an earthquake that travels through the Earth.
14. Cover the top of the gelatin with plastic wrap. Place a small cup filled with water on the gelatin and tap the pan.

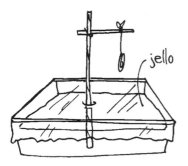

15. Let the students build some structures with sugar cubes or blocks on the gelatin and tap the pan. What happens to the structures? Construct a street light pole by tying two straws together and hanging a paper clip from the cross arm. Stick the pole into the gelatin and watch to see what happens when the gelatin "quakes."
16. Challenge the students to build a structure that is less likely to collapse when the gelatin shakes and explain what it looks like.
17. Tell the students to draw on the top half of the recording sheet *Earthquake Model* what the structures looked like before the shaking. On the bottom of the page, invite them to illustrate what the structures look like after the shaking.

Part 3
Safety
18. Tell the students that just like their gelatin model, an earthquake can shake, damage, or destroy buildings and freeway structures. (You may want to inform them that cracks in the Earth generally *do not* open up and swallow people, homes, and animals.)

19. Discuss what happened to the structures on the two earthquake models. Ask if the students feel that the classroom or their homes could be damaged in a large earthquake. Tell them the way to protect themselves is to be prepared when an earthquake happens.

20. Show the students the picture of things that could *possibly* happen during an earthquake and discuss the dangerous areas of a room. (See *Earthquake.*)

21. Give each student a copy of the *Earthquake Safety* sheet. Discuss the things that they and their families can do to prepare themselves *before* an earthquake happens.

22. For earthquake safety *during* an quake, show them the drop and cover method of protection (see sheet). Have the students go through a drill in the classroom. Instruct the students to turn away from the windows and crouch under a desk or table, protecting their head and neck with their hands. They should hold onto the legs of the desk or table and move along with it if it moves. Give the students the recording sheet and tell them to draw what they should do in an earthquake.

23. Explain to the students that if they are in a structure or area that has been damaged in an earthquake, they need to follow the instructions in the *after* column. Stress that they need to get adult help and to follow the directions they are given.

Discussion

1. What causes an earthquake?
2. Describe what you felt when you pressed your hands together and slid one.
3. Explain how you think that pressing your hands together and sliding your hands shows how an earthquake moves.
4. Explain how the Earth can quake and not cause damage to buildings.
5. Explain what you would do if you were in an earthquake at school... at home.
6. What caused the building in the *Earthquake Gelatin model* to fall down? [the shaking of the table]
7. What are some of the ways that earthquakes can be different from each other?
8. Have you ever felt the Earth shake from something other than an earthquake? What caused the shaking? [a large truck passing by] How do you think this is like an earthquake? How is it different?

Extension

Have students try to build structures that won't be damaged by the quaking of the *Earthquake Gelatin.*

Curriculum Correlation

Social Studies
1. Discuss the effects of earthquake damage on property, people, and communities.
2. Research the "Ring of Fire" around the Pacific Ocean.
3. Have students research the history of seismographs.

Language Arts
1. Have the students research earthquake legends from around the world.

Home Link

Have the students find the safest places in their homes to be during an earthquake. Encourage them to work with their families to identify the things that could fall or be broken during the shaking.

Earthquake Model

Observe earthquake models. Pick one to draw.
Tell about what you see happen.

Before

After

Earthquake Safety

Before

1. Plan what your family will need at home.

2. Store food, water, flashlight, batteries, first aid kit, radio with batteries.

3. Think of a safe place to be.

4. Plan where to meet if separated.

(((During)))

1. Stay calm. Turn away from windows.

2. Drop - get under a strong desk, table, stand in the doorway.

3. Cover - put both hands on the back of your neck. Tuck your head.

4. Hold legs of desk or table and move with it.

5. If outside, stay away from buildings.

After

1. Get help as needed.

2. Have an adult turn off the water and electricity if needed.

3. Listen to the radio for information.

4. Stay off the telephone except for emergency.

Earthquake

Name: _____

You are standing next to your school desk. The room starts to shake and you feel dizzy. Show what you would do in an earthquake.

Keep Safe!

I would _____
_____
_____
_____

Volcanoes

Topic
Volcanoes

Key Question
What are volcanoes and how do they erupt?

Focus
The students will be able to describe a volcano and how it erupts.

Guiding Documents
NCTM Standards
- *Develop the process of measuring and concepts related to units of measurement*

NSE Standards
- *The surface of the earth changes. Some changes are due to slow processes, such as erosion and weathering, and some changes are due to rapid process, such as landslides, volcanic eruptions, and earthquakes.*

Project 2061 Benchmarks
- *Change is something that happens to many things.*
- *Some changes are so slow or so fast that they are hard to see.*
- *A model of something is different from the real thing but can be used to learn something about the real thing.*

Science
Earth science
   volcanoes

Integrated Processes
Observing
Collecting and recording data
Applying
Communicating

Materials
Dirt
Sand
White vinegar (1/4 cup)
Baking soda (1/4 cup)
Red food coloring
Dishwashing soap (3 tablespoons)
Small jar
Large pan or flat box
Small bits of styrofoam (to represent rocks)
A half-empty tube of toothpaste
Pins
Newspapers
Cardboard tube
Drinking straws

Background Information
Volcanoes are a source of fascination and fear. Cultures from around the world have created stories to explain the causes of volcanic eruptions. The word *volcano* comes from Vulcano, an island near Sicily that was thought to be the home of Vulcan, the ancient Roman god of fire.

Magma is a mass of melted rock that originates in Earth's lower crust and upper mantle where temperatures may reach 700°C. The magma tends to rise because it is less dense than the surrounding rock. The magma, which is under a great deal of pressure, escapes through the weakest areas in the surface. These weak areas could be places where the crust is thin or cracked or along boundaries of Earth's plates. Magma that has reached Earth's surface is called *lava*. A *volcano* is an opening in the Earth's crust through which hot molten rock, gases, solid rock fragments, ashes, etc. from inside the Earth make their way to the surface. A volcano can take on different forms, but the one we usually think of is that of a mountain in the shape of a cone.

There are about 500 active volcanoes on Earth. Volcanoes are often located over areas where crustal plates meet, separate, or subduct (one plate is forced beneath another). Volcanoes that erupt where the plates are separating tend to be much gentler than those where the plates collide. As the plates pull apart, magma wells up to seal the gap between the plates.

There are or have been volcanoes in almost every part of the world. Although volcanoes may seem quite destructive, they are also considered quite beneficial. Volcanic ash is full of nutrients that enrich the soil. In many parts of the world, people risk living near volcanoes so they can tend farms on the fertile volcanic slopes. Steam and hot water from volcanoes are used for heating houses and for generating electricity in many parts of the world.

Management
1. Collect newspapers to cover the work area.
2. Put the baking soda in a small jar. (If desired, add small pieces of styrofoam to represent rocks.)
3. Mix the vinegar, soap, and food coloring in a jar. Make a cardboard tube (or use a toilet paper roll) and fit it around the jar. Cut holes in the tube for drinking straws or small plastic tubing which represent vents.
4. The reaction of vinegar and baking soda is used in the simulation of a volcanic eruption. When vinegar and soda are mixed together, a chemical reaction will occur. This chemical reaction will produce carbon dioxide gas bubbles. These bubbles will rise to the top of the volcano and spill over the sides. The soap is added for a more frothy volcano.

Procedure

Part 1

What is a volcano and what is inside?

1. Ask the students to describe what they think a volcano looks like. Have them discuss what they know about volcanoes. Explain that the magma (rock that has melted because of extreme pressure) collects under the surface in a magma chamber. When magma from the magma chamber reaches the surface, the volcano is said to erupt. Tell the students that they are going to use a tube of toothpaste to see how pressure can make a volcano erupt.

2. Ask a student to hold the tube of toothpaste in his/her hands. Have the student gently press on the capped tube and describe what the toothpaste is doing. Relate this to the way the pressure in the pools of magma inside the Earth moves the molten rock. Let several students squeeze the tube of toothpaste.

3. Ask a student to push against the tube to force all the contents of the toothpaste tube toward the capped end. Tell the students that the tube represents the mantle of the Earth and the contents of the tube represent the liquid magma under the mantle. Direct the student to continue to press on the tube while another student makes a pin hole in the toothpaste tube. Ask the students how this is like a volcano that is erupting.

Part 2

Making and erupting a volcano

4. Make the volcano model by mixing sand, dirt, and water. Use the proportion of three parts sand to one part dirt.

5. Cover a large area with newspapers. Place the jar with the baking soda in a flat pan or shallow box. Set the box in the middle of the newspaper-covered area. Fit the cardboard tube neatly around the top of the small jar. Fit drinking straws or small plastic tubes into the holes of the tube. Mold the sand and dirt around this structure to represent a volcano. Leave the top open around the cardboard tube and openings around the plastic tubes.

6. When it is time for an eruption, pour the vinegar mixture down the tube and into the jar. Stand back and watch.

7. Ask students to describe their observations and tell how it is like a real volcano that is erupting.

8. Ask students to explain how volcanoes change the Earth.

Discussion

1. What did the vinegar and baking soda mixture represent?

2. Describe what happened to the liquid that came out of the volcano?

3. What do we call the material that comes out of a real volcano?

4. What caused our model of a volcano to erupt?

5. What causes a real volcano to erupt?

Extensions

1. Have students do research on the location of volcanoes of the world, where they are, when they erupted, how much damage have they done, and what has happened to the land around them.

2. Do research to find what materials are erupted from a volcano and how they enrich the life of the people nearby.

3. Research how pumice (a volcanic material) is used in cleaning and polishing products.

Curriculum Correlation

Social Studies

On a map, find areas of high volcanic activity. Discuss whether there is any pattern to their location? (The areas are located along the plate boundaries. Students may also notice that they are often located where there are mountains.)

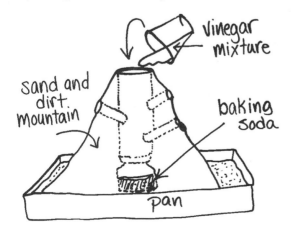

Volcano Model

You will need: dirt, baking soda, cardboard or
sand, vinegar, small jar
water, food coloring, pan
straws

Do this:

1. Put sand in a pan or small box.

2. Make a cardboard tube and fit it around the top of a small jar.

3. Cut 2 or 3 holes in the tube. Put plastic straws or small tubes in the holes.

4. Mix sand, dirt and water. Mold it around the jar on the tube to make a mountain.

5. Put baking soda in the jar. Mix vinegar, food coloring and dishwashing liquid in a cup.

6. Eruption Time! Pour the vinegar mixture into the jar and tube. Stand back and observe.

7. What happened?

vinegar + soap + food coloring

sand + dirt + water

baking soda in jar

pan

_____

_____

_____

89

Eruption

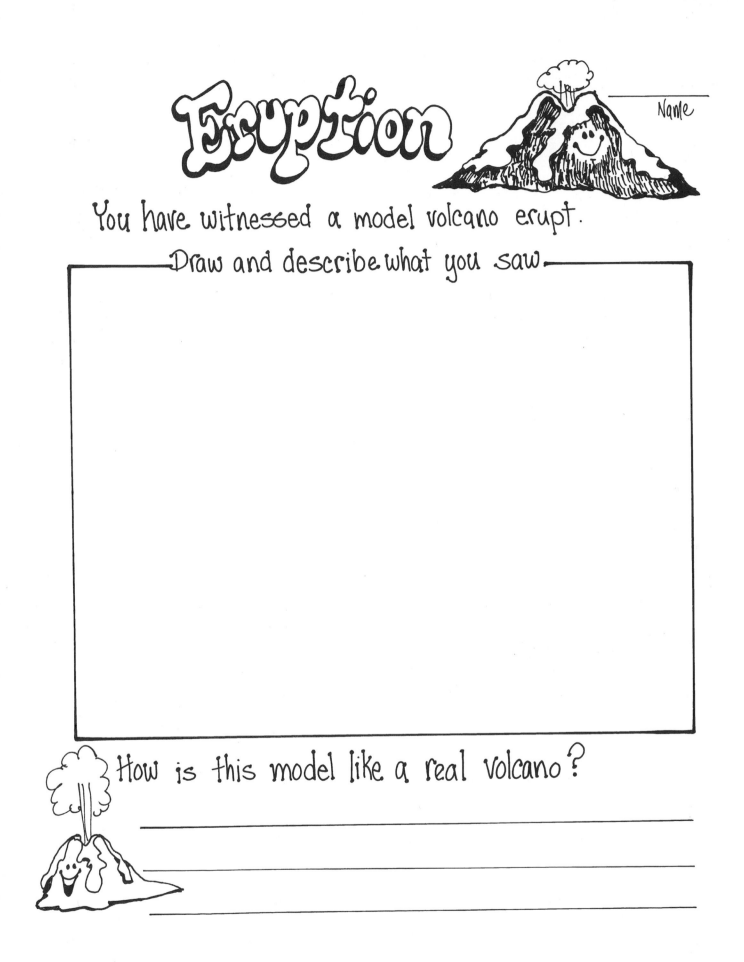

Name _____

You have witnessed a model volcano erupt.

Draw and describe what you saw

How is this model like a real volcano?

_____

_____

_____

Hydrosphere

Background Information

The study of water is a very important part of Earth Science. Water is a major architect of our landscape; it is an agent of erosion and essential to weathering both as a solvent and as a transport agent.

Water is vital to all life on Earth. Even the hardiest plants and animals of the desert need some water. Our bodies require water for proper functioning. Water is used in enormous amounts for agriculture, industry, and for many urban needs.

Water is a colorless, odorless, tasteless liquid. It is naturally found in three different states: liquid, gas, and solid. The state depends on the water's temperature. When water is cooled below its freezing point, it becomes solid. When water is heated, it can evaporate into the air as a gas. When it is above the freezing point and below the boiling point, it is liquid.

Earth is known as the water planet. There is water all around you—in the air above you, in the ground below you, and in the oceans. Water covers over 70% of the Earth's surface. While we speak of four different oceans, they are actually one vast body of water. The oceans contain 97% of the water on our planet. However, ocean water contains sodium chloride (salt) so its use by humans and most plants is limited. Only 3% of the water is fresh, and most of this fresh water is locked up in underground aquifers or in the icecaps at the poles.

Water is constantly being recycled through the oceans, the atmosphere, and the land. There is a continuous circulation as water moves through its cycle of accumulation, evaporation, condensation, and precipitation.

Water on the surface of the Earth and in the oceans where it has accumulated is evaporated by the sun's energy. The liquid becomes water vapor and with the right conditions of temperature and humidity, water vapor condenses to tiny droplets that form clouds and eventually falls as precipitation from the sky in the form of rain and snow.

Once the precipitation occurs, 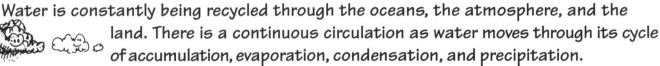 the water may soak into the ground, runoff and accumulate into lakes and oceans, or evaporate back into the atmosphere. The amount of water on our planet is the same today as millions of years ago.

Where is Water?

Topic
Water on Earth

Key Question
Where is water found on the Earth and how do we use that water?

Focus
Students will identify the places water is found.

Guiding Documents
NSE Standards
- *Earth materials are solid rocks and soils, water, and the gases of the atmosphere. The varied materials have different physical and chemical properties, which make them useful in different way, for example, as building materials, as sources of fuel, or for growing the plants we use as food. Earth materials provide many of the resources that humans use.*

Project 2061 Benchmarks
- *People need water, food, air, waste removal, and a particular range of temperatures in their environment, just as other animals do.*
- *Describing things as accurately as possible is important in science because it enables people to compare their observations with those of others.*
- *A model of something is different from the real thing but can be used to learn something about the real thing.*

Science
Earth science
 water

Integrated Processes
Observing
Inferring
Comparing and contrasting
Communicating

Materials
Globe
Large map of the U.S.
Sticky notes

Background Information
The Earth is often referred to as the water planet. Almost three-fourths of Earth's surface is covered with water. To help students become aware of this, this activity has them reflect upon the water in their immediate environment and then make applications to a larger area using a map and globe.

Most of the Earth's water is in the oceans. The oceans are made of salt water because water has run over the land for millions of years and brought minerals to the oceans. As the water evaporates, the minerals are left behind. Fresh water is found in lakes, rivers, streams, and ponds. Most of the Earth's fresh water is locked in the icecaps at the North and South Poles.

Procedure
1. Have the students think of all the places where water can be found and record using pictures and/or words on the recording sheet *Where is Water?* (Don't be surprised when they say the drinking fountain, the sink, etc. Accept these answers and continue probing if necessary.)
2. Ask the students to name any large bodies of water near them. Discuss the differences between the various bodies of water such as ocean, lake, stream, river, bay, pond, etc. Use pictures from travel ads and brochures or the illustrations provided.
3. Use the pictures provided to make a large chart. Ask the students to contribute words or phrases that describe each body of water (pond, lake, ocean, river, icebergs, puddle). Record their responses on the chart.

4. Invite students to make analogies about each body of water, such as
   A pond **is like** a giant puddle.
   An ocean **is like** a salty lake, so big you can't see the other side.
   A stream **is like** a tiny river.
   An iceberg **is like** a mountain of ice floating in the sea.
5. Hold a globe in front of the class. Ask them to make observations about the model of our Earth. Help them to see that all bodies of water are blue and there is much more blue than there is land showing on a globe. Point out that all the seas and oceans are connected, the water moves all over the world. Invite a student to trace his/her finger around the

globe pretending it is a ship that must travel all around the world. Let several students repeat the process to reinforce the idea that the oceans are all connected.

6. Brainstorm ways that we use water. Make a class chart of the students ideas. Begin with uses of water in the classroom, then at school, at home, in their neighborhood, city, state, country, and world.

7. Use a large map of the United States. Identify the areas of water on the map. Use sticky notes and have students draw pictures of people using the different bodies of water, then place them on the map. Only oceans, rivers, and large lakes will be marked large enough for you to see.

8. Have the students return to their recording sheet *Where is Water?* and add new illustrations or descriptions of what they have learned.

Discussion
1. Where did you learn that water is located?
2. What were some of the words used that described a river?
3. Think of the words you used to describe an ocean and a lake. How are these two bodies of water alike and how are they different?
4. Looking at the globe, why do you think the Earth is called the water planet?
5. How have you used water today?
6. From where did the water you used come?

Extension
Make a wave in a bottle. Use a two-liter plastic bottle with a lid. Put in one liter of water with a small amount of blue food coloring. Put in one-half liter of mineral oil and secure the cap. The water and oil will not mix. Tilt the bottle back and forth to make waves. Observe the wave action. Place a small floating object inside and observe how the waves affect it.

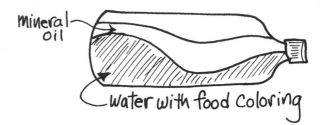

Curriculum Correlation
Art
Have the students use water colors to paint a picture of themselves swimming with water animals.

Language Arts
Make alliterations for each body of water
1. river – racing, raging, roaring
2. pond – puddly, peaceful, pool
3. stream – still, streaked, stony
4. lake – large, level, lasting
5. iceberg – immense, icy, ice cold

lake

icebergs

ocean

pond

stream

river

Where is Water Found?

Think of all the places you have seen water. List and draw.

Name: _____

What Makes Rain?

Topic
Rain

Key Question
What makes the rain?

Focus
The water cycle

Guiding Documents
NSE Standards
- *Ask a question about objects, organisms, and events in the environment.*

Project 2061 Benchmarks
- *The sun warms the land, air, and water.*
- *Water can be a liquid or a solid and can be made to go back and forth from one form to the other. If water is turned into ice and then the ice is allowed to melt, the amount of water is the same as it was before freezing.*
- *Water left in an open container disappears, but water in a closed container does not disappear.*

Science
Earth science
    meteorology

Integrated Processes
Observing
Classifying
Predicting
Collecting and recording data
Applying
Communicating

Materials
Glass or plastic jar with lid
3 small jars, one with lid
Water
Ice cubes
Tin cans
Self-closing gallon size plastic bags

Background Information
The water cycle is the process by which water leaves the Earth's surface through evaporation, rises into the air, condenses and falls as precipitation back to the Earth's surface. A cycle is something that repeats itself over and over again.

Amazingly there is as much water on Earth today as there was millions of years ago. Due to the cycling of water through the stages of evaporation, condensation, precipitation, and accumulation, the same water appears over and over again. Eventually every drop of the Earth's water goes through every stage of the water cycle.

Water as a gas is called water vapor and makes up about two to four percent of the air. Water vapor is added to the air by the process called *evaporation*. When water evaporates, it changes from the liquid to the gas state.

Condensation occurs when water vapor changes from a gas to a liquid. When conditions are right, the liquid falls as *precipitation*. Precipitation can take several forms: rain, snow, sleet, and hail. Most precipitation falls on the ocean. That which falls on land may run off into streams, rivers, etc., or it soaks into the land. Water *accumulates* in oceans and lakes where it continues the cycle by evaporating.

Management
1. The water needs to be very hot. It is recommended that the teacher pour the water. **Remind the students to be very careful with the hot water.**
2. Make transparencies of the two pages, *Rain and the Water Cycle* showing the mountains and lake and the following page with the arrows showing a water cycle. The transparencies will be placed one on top of the other.

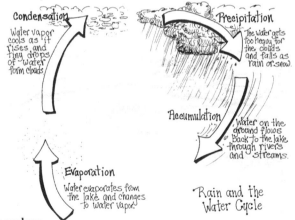

Rain and the Water Cycle

Procedure
1. Ask the students what they know about rain. How does it get to Earth? Where do the clouds come from? How does water get into the clouds? As a pre-assessment, have the students draw a picture explaining rain.
2. Ask students if the agree or disagree with this statement: Water is in the air at all times. Spend some time discussing why they believe the way they do. Tell them that they are going to do an activity that may help them understand that water is in the air.
3. Place three small jars in the window. In one jar, put an ice cube, in the second and third jar, put equal amounts of water but close the third jar with a lid. Have the students make observations over the next

96   © 1996 AIMS Education Foundation

few days. They should notice that the ice melts to water then disappears (evaporates), the water in jar two evaporates, and the water in jar three remains. Discuss and record their ideas on the sheet *The Three Jars*. Be sure to keep reinforcing the idea that the water evaporates—it goes somewhere.

4. Give each group of students a small can of ice water to observe. Have them feel the outside of the can when the water is first poured. It should be dry. After a few minutes there will be condensation on the outside that they can see and feel. Ask them where they think the water came from. [the air] If students believe that the water came through the can, repeat the experience and add food coloring to the water. The water on the outside will not be colored, it condenses from the surrounding air which is warmer. Have the students wipe the can with a white paper towel and they will see that it has no color. You can also remind the students that they will often see water on the outside of their milk cartons at lunch. Ask them where they think that water came from.

5. Invite a student to get very close to the chalkboard, take a deep breath, and exhale on the board. Ask the students to explain what happened. [The breath, which has moisture, condensed on the chalkboard and disappeared when the moisture, evaporated. The moisture went into the air.]

6. Entice the students with the question: Do you think we can make rain in a jar? Allow time for discussion.

7. Tell them that you have a recipe for making rain in a jar. Pour about five centimeters of very warm water into the jar. Place the lid upside down over the mouth of the jar. Allow it to sit for a few minutes. Then place ice cubes in the lid. Darken the room and shine a light through the jar. (Use a flashlight or use the light from the overhead projector.)

8. Ask the students to describe what they see happening in the jar and why they think that this is happening. They should be able to see wisps of clouds forming and moving upwards. "Raindrops" will form on the lid and fall back to the water. Ask students if they know how this activity relates to nature making rain. If necessary explain that the sun heats the Earth, making water evaporate into the air. High above us the air is much cooler and the water condenses into clouds and falls back to the Earth as rain or snow.

9. Show the transparencies of the water cycle. Use this to tell the story of how what happened in the jar is like what happens in the real world. Allow students to use their own words. If they are developmentally ready for the vocabulary, it can be used.

10. Put warm water in a self-closing plastic bag and seal tightly. Be sure to leave some air in the bag. Lay the bag on its side and observe. Have students,

one at a time, place a hand on top of the bag. Their hands are colder than the bag and water will condense in the shape of their hand on the top of the bag. Give them an ice cube to rub slowly across the bag. The cube will create streaks of cloud wherever it touches.

11. Ask students to look at the picture they drew at the beginning of the lesson. Have them discuss whether they now agree or disagree with what they drew to show where rain comes from.

Discussion

1. Explain what went on in the three jars. Where did the water in the two jars go?

2. What do you feel on the outside of the can with the ice water inside? Where do you think the moisture came from?

3. What do you observe rising from the hot water in the clear jar?

4. What do the drops that fall back into the water represent?

5. Explain what happens when water evaporates from a water body on Earth? Where does it go? Does it stay there?

6. Can we see the water in the air? Explain. [Sometimes—when it rains or snows or its foggy, we can see clouds. Other times its invisible.]

Extensions

1. Do the activity *Drying on the Line* (*AIMS*, Volume 9, Number 7).

2. Extend the idea that there is water in the air by making frost. Fill a clean metal can with ice, then add rock salt. Observe the can while discussing the possibilities of what might happen. Slowly, frost will form on the outside of the can. The temperature of the salt water in the can is below the freezing temperature of water. The water that has condensed on the outside of the can freezes forming the frost.

3. Another way to make rain without using ice is to heat water in a pan until it is boiling. Hold a clear lid over the steam, water will condense on the lid and fall back to the pan like rain.

Curriculum Correlation

Math

Place a wide-mouth jar of water in a sunny spot on the window ledge. Mark the level of water with a marker or a piece of masking tape. Mark and measure the level each week to keep track of evaporation. Have students predict the day they think all the water will be gone.

Literature

Brandt, Keith. *What Makes it Rain?* Troll. Mahwah, NJ. 1982.

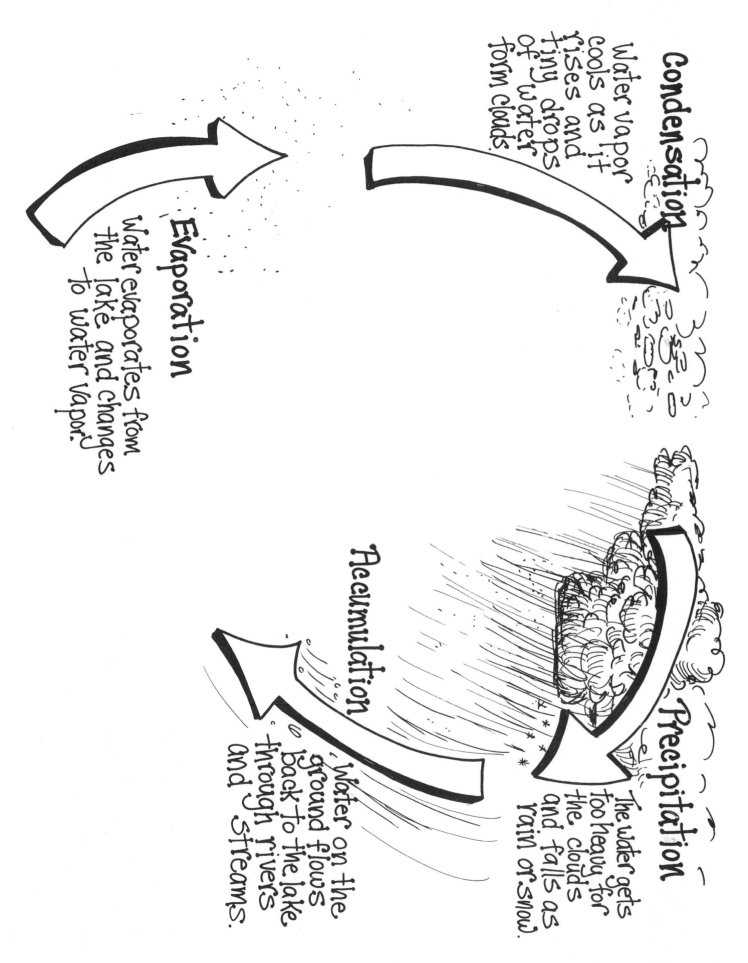

Condensation
Water vapor cools as it rises and tiny drops of water form clouds.

Precipitation
The water gets too heavy for the clouds and falls as rain or snow.

Accumulation
Water on the ground flows back to the lake through rivers and streams.

Evaporation
Water evaporates from the lake and changes to water vapor.

Rain and the Water Cycle

The Three Jars

Name_____

Observe the 3 jars on the window ledge.
Draw and explain what you see happen.

Open jar	
Open jar	
Closed jar	

What Makes Rain?

1. Put warm water in a jar.
2. Put ice in the lid.
3. Draw what you see happen.
4. Why do you think this happened? Please explain.

Name _____

upside down lid

A Disappearing Act

Topic
Evaporation of water

Key Question
What happens to moisture on the playground and on the chalkboard?

Focus
The students will see the results of water disappearing (evaporating) into the atmosphere.

Guiding Documents
NCTM Standards
- *Develop the process of measuring and concepts related to units of measurement*
- *Make and use measurements in problems and everyday situations*

NSE Standards
- *Ask a question about objects, organisms, and events in the environment.*
- *Employ simple equipment and tools to gather data and extend the senses.*

Project 2061 Benchmarks
- *People can often learn about things around them by just observing those things carefully, but sometimes they can learn more by doing something to the things and noting what happens.*
- *Water left in an open container disappears, but water in a closed container does not disappear.*
- *The sun warms the land, air, and water.*
- *Tools such as thermometers, magnifiers, rulers, or balances often give more information about things than can be obtained by just observing things without their help.*

Math
Measurement
    time
Estimation

Science
Earth science
    water cycle
        evaporation

Integrated Processes
Observing
Comparing and contrasting
Collecting and recording data
Predicting
Generalizing

Materials
Balance
Two identical jars with lids
Water
Chalk
Cloth or paper towel
Stopwatch, optional
Ruler (see *Management 4*)

Background Information
Evaporation is the process by which a liquid changes to a gas. When water evaporates, it changes to an invisible, odorless gas called water vapor. Changes in states of matter require a transfer of energy. Energy from the sun causes this evaporation.

Students are not able to witness the evaporative process; they can, however, observe the results of evaporation. In this activity, students will observe puddles on the playground and notice that the water disappears; it goes somewhere. They will also observe the disappearance of moisture on the chalkboard or table top. They will collect data by determining the time it takes for the moisture to evaporate.

Management
1. This activity has three parts. *Part 1* has students observe puddles on the playground. *Part 2* takes place indoors with students observing the disappearance of the moisture from the chalkboard and from their hands. *Part 3* lets students observe what happens to a balance as water evaporates.
2. The puddles that are being observed on the playground need to be in an area that will not let the water soak into the surface. A concrete sidewalk or paved play area work well.
3. If the weather has not cooperated with rain, make some puddles by pouring water into a depression in the surface being observed.
4. If students do not know how to use rulers, substitute non-customary units such as Unifix cubes. String can also be cut to represent the measure of the puddle.

Procedure
Part 1
1. Take the students out on the playground after a rain to observe the puddles. Have the students use chalk to trace around the puddles. Watch the puddles for a period of time. Have the students draw a "before" view of the puddle.
2. Have students measure the diameter of the puddle. Many puddles will not be symmetrical, so have students determine how they are going to measure them and include their methods in their illustrations.

3. Periodically check the puddles to see how much evaporation has occurred. When you feel that it is substantial enough to make an impact on the students, invite them back outside to illustrate and measure their puddles once again. Have the students measure and draw an "after" view of their puddle.
4. Have the students explain what they think happened to the water in the puddles.

Part 2

5. Tell students that they are going to once again watch water disappear. Wipe a damp cloth across the chalkboard. Ask students to explain what they think is happening.

6. Again wipe a small area on the chalkboard, countertop, or classroom floor. Draw a circle around the wet area. Watch the moisture disappear. Have the students count, or time with a stopwatch, how long it takes the water to disappear. Discuss with the students what is happening and where the moisture may be going.
7. Have the students wash their hands but not dry them. Have them time how long it takes for their hands to dry? Ask them to explain what things might cause their hands to dry more rapidly... more slowly.
8. Discuss how the disappearance of water from the chalkboard and from their hands is like the water in the puddle.
9. Ask students to brainstorm other things they could do to observe that water disappears.

Part 3

10. Fill two identical containers with the same amount of water, put lids on both jars. Place them in the balance and make certain that it is equalized. Remove the lid from one jar and put in the pan so the mass will remain the same. Leave the water and jars intact and observe over a period of several days.

11. Watch what happens to the balance and the water level in the jars. Discuss with the students why the balance is tipping. Have the students draw and write about what they see happening on the *Covered and Uncovered* recording sheet.
12. Ask students to apply the disappearance of water from the jars to the experience with the puddles.

Discussion

1. When a puddle dries up, where does the water go?
2. What happened to the wet place on the chalkboard? Where did the water go?
3. How can there be water in the air? [The drops are so tiny we cannot see them.]
4. Is there water in the air that we can see? [clouds, fog]
5. Could you see the water evaporate? Why not?
6. What would happen if there were twice as much moisture in the spot on the chalkboard, or floor? Would it take twice as long to evaporate?

Extensions

1. *Parts 1* and *2* of this activity could be repeated on a very humid day and on a dry, sunny day. Results should be compared and discussed.
2. Fill several different-sized containers about one-half full of water. Try to pick containers that have different-sized openings. With a black marking pen, mark the level of the water at the beginning. Watch the level of the water for several days, mark the level of the water daily. Does the size of the opening make any difference in the evaporation of the water?
3. Do *Drying of the Line* (AIMS, Volume 9, Number 7).

Curriculum Correlation

Language Arts

Cole, Joanna. *The Magic School Bus at the Waterworks*. Scholastic, Inc. NY. 1986.

Peters, Lisa Westberg. *Water's Way*. Arcade Publishing. NY. 1991.

A Disappearing Act

1. Go outside. Observe some puddles of water. Come back later.

Draw the puddle

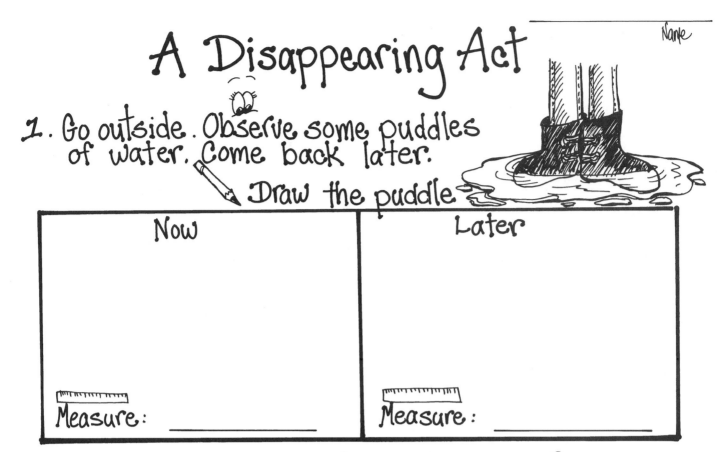

Now	Later
Measure: _____	Measure: _____

What happened to the water in the puddle?

2. Wipe an area with something wet. How long will it take for the water to disappear?

Try 1 :   I think _____. I counted _____.

Try 2 :   I think _____. I counted _____.

Try 3 :   I think _____. I counted _____.

3. Where do you think the water goes?

Covered and Uncovered

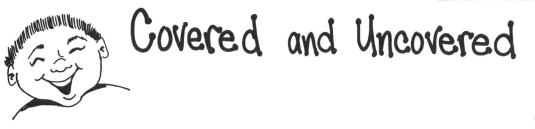

1. Make 2 jars of water just alike.

2. Put them in a balance. Make sure both sides are equal.

3. Take the lid off one jar.

4. Leave alone for several days.
   What do you think will happen?

5. Draw and write about what happened.

105

Water to Ice to Water

Topic
Water in liquid and solid form

Key Question
What happens to water when it freezes?

Focus
The students will discover that water expands as it freezes and that it will float.

Guiding Documents
NCTM Standards
- *Make and use measurements in problems and everyday situations*

NSE Standards
- *Materials can exist in different states—solids, liquid, and gas. Some common materials, such as water, can be changed from one state to another by heating and cooling.*
- *Objects have many observable properties, including size, weight, shape, color, temperature, and the ability to react with other substances. Those properties can be measured using tools, such as rulers, balances, and thermometers.*

Project 2061 Benchmarks
- *People can often learn about things around them by just observing those things carefully, but sometimes they can learn more by doing something to the things and noting what happens.*
- *Describing things as accurately as possible is important in science because it enables people to compare their observations with those of others.*
- *Water can be liquid or a solid and can be made to go back and forth from one form to the other. If water is turned into ice and then the ice is allowed to melt, the amount of water is the same as it was before freezing.*

Math
Measuring

Science
Earth science
  ice and water

Integrated Processes
Observing
Comparing and contrasting
Collecting and recording data

Materials
Plastic cups
Water
Masking tape
Plastic wrap
Containers of water
Rulers (see *Management*)

Background Information
Water is the only substance on Earth that is naturally present in three different states: as a liquid, a solid (ice), and a gas (water vapor).

Water has a very unusual property compared to other liquids. The volume of most liquids decreases as they freeze. As water cools, it contracts to a point but then begins to expand as it reaches its freezing point. Because of this expansion, ice floats on water. As a result, rivers and lakes freeze from the top down rather than from the bottom up.

Most of Earth's fresh water is stored in glacial ice. So far, it has not proved practical to transport it to arid regions for drinking, irrigation, or industry use.

In this activity, students will compare and contrast liquid water and ice. They will then discover that ice floats which can lead to the Earth Science application of icebergs.

Management
1. In *Part 1*, each group should have a cup of water and a cup that has water to be frozen. The containers of water used in *Part 2* should be large enough to hold water and the ice from the cups in *Part 1*.
2. For the measurement of the height of the water, customary or non-customary units (i.e., Unifix cubes or string) may be used.
3. You will need to have a freezer available, or do this in the winter when the cups of water can be placed outside to freeze.

Procedure
Part 1
1. Begin the activity by asking students where and when they see ice in nature. Ask them to explain the conditions that create ice in nature. [There must be water and it must be cold.]
2. Inform students that they are going to look at what happens when water freezes. Give each group of students two cups with equal amounts of water in them. Ask them what they think will happen if one of the cups of water is left in the freezer overnight. Have each group put their names on masking tape and then use the tape to mark the water levels in their cups. Cover the cups with plastic wrap to prevent any evaporation.
3. The next day distribute to each group their cup of ice and their cup of water. Discuss changes in the water when it became ice. If students do not

mention the height variable, direct them to notice the level of ice in comparison to where the tape marked the level of water. Have them measure the height of the ice and mark it with another piece of masking tape.

4. Invite the students to compare the contents of the two cups. Ask students for words that describe the water and the ice. Make one list for water and one for ice. Encourage the students with questions like: (a) Which feels colder, the water or the ice? (b) Can you hold the ice in your hand? What about the water? (c) Can you pour the water? Can you pour the ice? (d) What sound do they make when you pour them out? (e) Can you see through the water? ... the ice? Have them record the comparisons on the sheet *Water to Ice to Water*.

5. Compare and contrast the lists of physical characteristics.

6. Let the ice in the cups melt. Have students look at the height of the water. They should see that it is the same level they marked before they froze it. Ask them what they think will happen to the height if they put the cups back in the freezer. Record their predictions on the chalkboard.

7. Freeze the water and have students again observe the height of the ice in the cups.

Part 2
Changing ice to water to ice

8. Using the cups of ice, ask the students what will happen to the ice when they put it into a container of water. If they do not include ideas about melting and floating, ask questions to direct their thoughts to these types of predictions.

9. Have the students carefully put the ice into the containers of water and discuss what happens.

10. Direct them to push the ice to the bottom of the container and observe what happens when they release it. Ask them if they can think of any way to keep the ice on the bottom of the container. Allow them time to try their ideas, but always bring them back to the fact that the ice, by itself, will float.

11. Ask students to illustrate the ice in their container of water.

12. Leave the ice in the containers of water so that the students can observe that it eventually melts, turning back into its liquid form.

Discussion

1. How are water and ice alike?... different?

2. What did you notice about the height of the ice as compared to the height of the water? What happened to the height when the ice melted? [It was the same as the water level before it was frozen.]

3. Explain what happened to the ice when you put it in the water.

4. Were you able to keep the ice from floating when you weren't holding it down or using something else to hold it down? [No, it always floated.]

5. In the ocean where it is very cold, icebergs are found floating in the water. What do you think an iceberg is? How is it like the ice floating in your cup?

6. When lakes and ponds freeze over during the winter, where do you think the ice is found, at the bottom or at the top? [At the top] Explain why you think this way. [It is like the ice in our cup, it floats.]

7. What do you think will eventually happen to your ice that is in the container of water? Why does it melt?

8. What do you think eventually happens to the glaciers that float in warmer water?

9. What happens to the frozen lakes and ponds when spring comes? Why?

Extension

Do the activity *Melt an Ice Cube* from the AIMS publication *Primarily Physics*.

Curriculum Correlation
Social Studies

1. Look at the globe and observe where the icecaps of the Earth are.

2. Make a large ice cube in a plastic bag. Place it in an aquarium of salt water so students can see that it floats. Explain how this is like an iceberg in the ocean. It floats, but most of the iceberg is hidden beneath the water where it is a hazard for boats.

Water to Ice to Water

1. Make 2 cups of water. Measure and mark the levels.
2. Freeze one of the cups.
3. Record your observations.

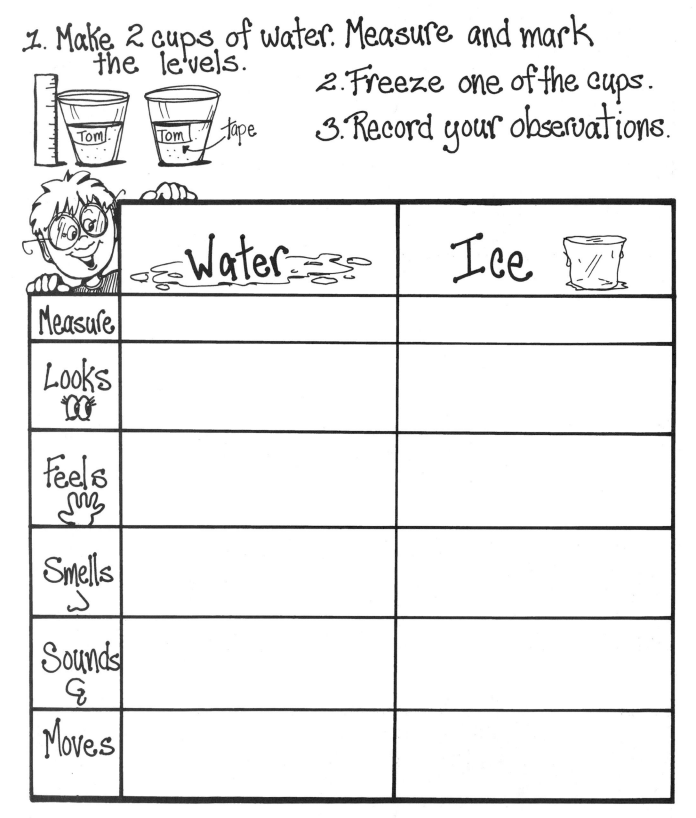

	Water	Ice
Measure		
Looks		
Feels		
Smells		
Sounds		
Moves		

Ice in Water

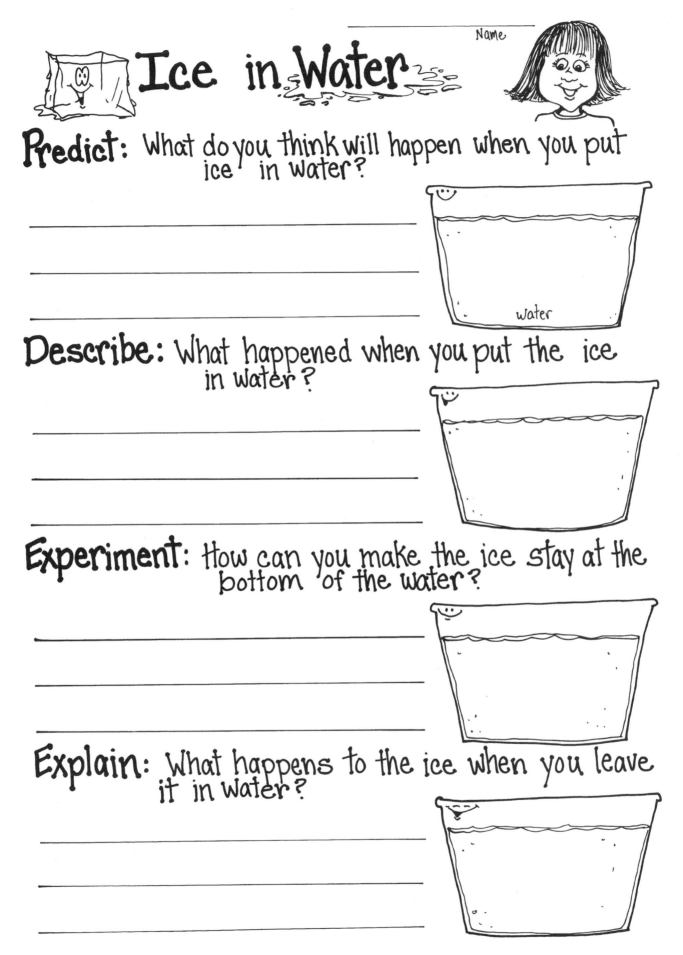

Name _____

Predict: What do you think will happen when you put ice in water?

_____
_____
_____

water

Describe: What happened when you put the ice in water?

_____
_____

Experiment: How can you make the ice stay at the bottom of the water?

_____
_____

Explain: What happens to the ice when you leave it in water?

_____
_____

Atmosphere

Background Information

Earth is surrounded by an ocean of air. It is like a blanket that helps protect us from the harmful rays of the sun, traps the warmth of the sun, and gives us oxygen to breathe. Air is one thing in our environment that is always around us. It is in the atmosphere, in the clouds, in water, in soil, and inside living things. Air is several gases mixed together, mostly nitrogen, with oxygen, water vapor, carbon dioxide, and other trace gases. All living things must have air in order to live.

Air exerts pressure. We do not normally notice this pressure because we are supported by equal pressure on all sides of our bodies and it is balanced by the pressure inside our bodies. Air pressure is caused by the pull of gravity on the air and the activity of the molecules in the gases (a result of air temperature). Air pressure decreases with height. A ride up the mountain or a trip by airplane lets you notice this difference in air pressure.

Wind is moving air. The atmosphere is always in motion because of the unequal heating of the Earth–warm air expands, becoming less dense, and rises, while cooler air is more dense–and because of varying pressure areas. Surface winds result from horizontal pressure differences. When the differences are greater, the winds will be stronger; with smaller differences, gentler winds will occur. Although the winds generally move from areas of high pressure to areas of low pressure, there is some deflection because of the Earth's rotation.

Winds are named after the direction from which they blow. A wind that blows from the northwest toward the southeast is called a northwest wind. Wind vanes are used to show the direction of wind. A wind vane is blown around by the wind so that it points into the direction from which the wind is coming. The speed or force of a wind can vary from a gentle breeze to a hurricane. The speed of the wind can be measured by an instrument known as an anemometer. Before modern anemometers were invented, the strength of the wind was judged from its effects on land objects (trees, flags) and on the surface of the water.

The atmosphere serves as a huge fresh-water reservoir where water exists in three states: solid, liquid, and gas. Clouds are formed from water that has evaporated from the Earth's surface. Clouds are a mass of small water droplets that float in the air. They are named according to their formations and the altitude at which they are formed. The three basic groups are cirrus, stratus, and cumulus.

Weather is the condition of the atmosphere at a certain time in regards to the amount of sunlight, precipitation, clouds, and range of temperature. Weather affects our lives in many ways. The type of clothing we wear depends largely on the weather. The agriculture that can occur in an area is largely dependent upon that area's weather. Businesses and transportation are affected by weather, good and bad.

Our atmosphere is amazing: Hurricanes, tornadoes, gentle breezes, sunsets, hail, lightning, fog, and thunderclouds all originate from four basic ingredients: air, water vapor, tiny suspended particles, and sunlight.

A Close Look at Air

Topic
Air

Key Question
What can we observe about air?

Focus
Air is all around us even though we can't see it.

Guiding Documents
NSE Standards
- *Ask a question about objects, organisms, and events in the environment.*
- *Communicate investigations and explanations.*

Project 2061 Benchmarks
- *Often you can find out about something big by studying just a small part of it.*
- *We can learn about things around us by just watching carefully or by doing something to them and seeing what happens.*
- *Describing things as accurately as possible is important in science because it enables people to compare their observations with those of others.*

Science
Earth science
   air

Integrated Processes
Observing
Contrasting and comparing
Communicating
Applying
Generalizing

Materials
Plastic zipper-type bags
Crackers

Background Information
Young learners often have trouble accepting that air exists because, when it is not polluted, it has no color, smell, or taste. One way to help them understand that air exists is to show them that air takes up space not taken up by other things. By filling a plastic bag or balloon with air, they can experience that "something" is there. Even though they can't see it or smell it, it still exists. Another way to help them understand that "something" is there is to let them feel it as they wave their arms, stand in front of a fan, or face the wind.

All living things must have air in order to live. Our Earth has air in its atmosphere, in the clouds, in water, in soil, even inside living things.

Management
1. Collect several things which can be used to demonstrate that air is present: a tire pump, an inflated balloon, a squeeze bottle, an electric fan, etc. Set these things out on a table to be used at the end of the lesson. Have students apply what they know about air to these objects.
2. If appropriate, introduce the word *invisible*.

Procedure
1. Have the students in the class take a deep breath and feel the air go into their lungs. Have them take another breath and breathe on their hands to feel the warmth and moisture from the air inside their bodies. Tell the students to take a deep breath and hold it for 20 seconds. Explain that their lungs may burn slightly because their bodies need air. Discuss with the students that we need air in order to live.
2. Discuss with the students their understanding of air: what it is, what they see, what they can do with it, and where it is. Record their ideas on a chart for later reading and review.

> **What is Air?**
> Air is what I breathe.
> Airplanes fly in air.
> The air moves in wind.
> Plants make oxygen.
> Animals need air.
> Air is in tires.

3. Show the students a flattened plastic bag. Open it and ask a student to come up, look inside, then tell the class what is in the bag. (The students will usually respond that the bag is empty.) Capture air in the bag and close the top so that air is trapped inside. Have the student press on the sides of the bag. Ask them why the bag doesn't collapse.
4. Give each pair of students a plastic bag. Have them "catch" some air in it and then close it tightly.
5. Have the students squeeze the trapped air and describe how it feels and looks.
6. Ask the students what will happen if they put a book on their bag. Again ask them why the bag didn't collapse.
7. Give each group one or two crackers. Tell them to unzip their bags, put the crackers inside, catch some air, and zip the bag  shut. Ask them what will happen when they put the book on the bag this time. [The crackers won't be crushed because the air in the bag takes up space.]

8. Have the students take the crackers out of the bag. Once again have them capture some air in the bag and smell it. Ask them if it has an odor. Direct the students to record their observations on the recording sheet *A Close Look at Air*.

9. Have the students stand up and swing their arms up and down to feel the air in the classroom. Ask them to notice the difference as they swing their arms at different rates. Give them a piece of cardboard to swing. Ask them what they feel. [air moving] Have them record what they observe on the recording sheet *A Close Look at Air*.

10. Turn off the lights and turn on the overhead projector (or a flashlight). Let the students observe the dust particles that are floating in the air. (If none can be seen, hit two chalkboard erasers together in front of the overhead.) Ask the students to describe what they see. Emphasize that they cannot see the air, but they can see things that float in the air.)

11. Use the patterns on the sheet *Observing Air* to make helicopter bunnies and elephants. Have students drop the completed helicopters and watch them as they fall. Hopefully they will notice that the blades are lifted by the air as they spin to the ground.

12. Hold a competition to see which helicopter stays in the air the longest. Have students count the seconds and compare results.

13. Generate a list on a chart labeled *Where is Air?* of where they can find air. Add to the chart as students discover more places that air is.

14. Have the students invent something that shows air is all around us. For example: kite, paper airplanes, wind toy, confetti, wind sock, parachute. Use the recording sheet *Air is All Around Us*.

Discussion

1. Why didn't the crackers in the bag break?
2. Can you feel air? Explain.
3. How can you show that air is around you?
4. Explain how your helicopters are like a real helicopter. How are they different?
5. Did the size of the ears of the rabbit and elephant make a difference in how your helicopter flew? Explain.

6. What can you do to the rabbit and elephant that will make it fly better? (Possible answers: Add a paper clip to the bottom for a little more weight. Drop it from a higher spot.)
7. How does your invention use air?
8. What did you learn about air?

Extensions

1. For a math lesson, have the students count the number of breaths they take in one minute and record on a chart. Challenge the students to find a way to determine how many breaths they take in 10 minutes, an hour, etc. Have them compare and contrast their totals and attempt to explain any differences. (see *Learning About Lungs*, AIMS magazine, Volume 10, No. 5)
2. Challenge the students to make the helicopters spin in the opposite direction.
3. Ask students to bring in objects that show that air takes up space and we can feel air or know in some way that air is everywhere.

Curriculum Correlation

Art

Do spatter painting by dropping tempera paint onto paper and carefully blowing the paint in different directions. Point out to students that they are using air instead of brushes to paint.

Language Arts

Brainstorm with the students a list of favorite smells. Have them write about those smells or write a poem about the smells. Often a class poem can be fun for all. Here is a sample poem.

I love the smell of flowers in spring.
I love the smell of bread baking.
I love the smells of the county fair.
I love the smell of clean fresh air.
I love the smell of grass just mowed.
I love the smell of gardens hoed.
Smells travel in the moving air.
My nose smells smells everywhere.

—*Sheryl Mercier*

A Close Look at Air

1. Catch some air in a plastic bag. Observe.
   Write and draw.

Look	Smell
Touch	Hear

2. Stand up and swing your arms slow and fast.
   What do you feel?

3. Hold a piece of cardboard in each hand. Swing
   your arms slow and fast.
   What do you feel?

Observing Air

1. Color and cut out.
2. Fold and tape sides back.
3. Put paper clip on the bottom.
4. Bend the blades.
5. Drop the helicopters and compare.

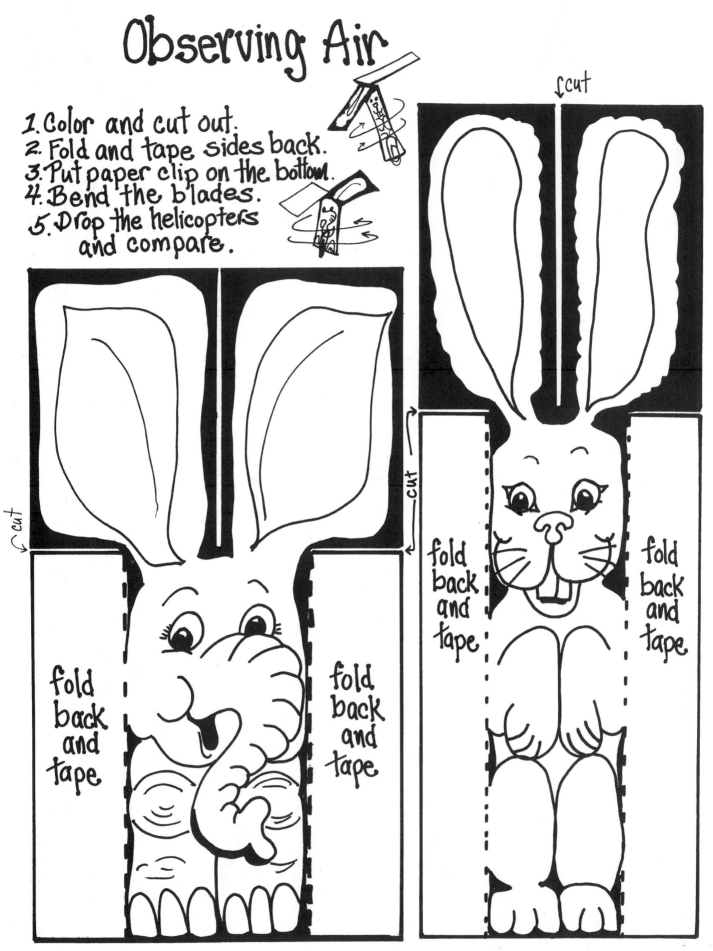

fold back and tape

fold back and tape

cut

cut

cut

fold back and tape

fold back and tape

114

Air is All Around Us

You are an inventor. Design an invention that shows air is around us. Use it for the class.

I call my invention _____

Here is a picture of what I invented.

[blank box]

My invention worked

Very well   O.K.   I need to start over

Air is There

Topic
Air

Key Question
How do we know air takes up space?

Focus
Air takes up space.

Guiding Documents
NSE Standards
- *Ask a question about objects, organisms, and events in the environment.*
- *Communicate investigations and explanations.*

Project 2061 Benchmarks
- *We can learn about things around us by just watching carefully or by doing something to them and seeing what happens.*
- *Most living things need water, food, and air.*
- *Tools such as thermometers, magnifiers, rulers, or balances often give more information about things than can be obtained by just observing things without their help.*

Science
Earth science
   atmosphere

Integrated Processes
Observing
Contrasting and comparing
Communicating
Predicting
Applying
Generalizing

Materials
Plastic gallon jug
Large plastic bag, not zipper-type
Piece of styrofoam
Food coloring
Toothpick
Clear plastic cups
Paper towels
Aquarium

Background Information
Air is a mixture of gases and like all gases, it takes up space. Non-polluted air is invisible. It is often difficult for young learners to understand that "something that is invisible" actually exists. Using the avenue that air takes up space, students will have multiple experiences in which they should begin to conceptualize this.

The air surrounding Earth is like a protective blanket that protects us from the harmful rays of the sun, helps trap warmth, and gives us oxygen to breathe. Air is in the atmosphere, in the clouds, in water, in soil, even inside living things.

All living things must have air in order to live. Humans use air for many reasons such as flying, parachuting, hang gliding, flying kites, inflating tires, lifting weights with hydraulics, and generating electricity to name a few.

Procedure
1. Tell the students that you are going to give them a riddle to solve. Point out that there are some rules to answering the riddle: Raise your hand as soon as you know the answer. Don't tell the answer until the riddle is finished.
   I am invisible.
   I am everywhere.
   I move easily.
   I surround the Earth.
   All living things need me.
   I can carry birds, clouds, and airplanes.
   I am in your lungs.
   What am I?
2. Discuss with the students that air is all around us and even if we don't see it, there are things we can do to find out that it is there. Ask them for suggestions of things that they could do.
3. Put water in an aquarium or large wide-mouth jar. Place a crumpled paper towel in the bottom of a clear plastic cup. Put the cup right-side up into the water. (The cup should float.) Push the cup so that water enters, soaks the paper towel, and the cup eventually sinks. Ask the students why the paper towel got wet. [Water got in the cup.]
4. Crumple a paper towel and put it in the bottom of another clear plastic cup. Tell the class that you are going to turn the cup upside-down and push it into the water. Ask them what they think will happen. Push the inverted cup to the bottom of the water. Bring it back out and show that the paper towel did not get wet. Ask them why they think the paper towel didn't get wet. [Air is in the cup so water couldn't get in.]
5. Add a few drops of food coloring to the water. Make a small boat using a piece of styrofoam for the hull, a toothpick for the mast, and a piece of paper for

the sail. Ask students to predict what will happen to the boat when you place a cup over it and push the cup to the bottom of the water.

6. Place a cup upside-down over the boat and carefully push the cup to the bottom of the water. Have the students observe and explain why they think the water does not enter the cup. [There is air inside the cup that keeps the water out.]

7. Tilt the cup a little so that some air bubbles escape. Ask the students what made the bubbles. Have them observe what has happened inside the cup. [The water level is higher.] Discuss why the water level is higher. Raise the cup to the water line and have students describe what they see.

8. Let students take turns placing the inverted cup and paper towel and the inverted cup and sail boat into the water. Make certain that students refer to the fact that air takes up space in the cup.

9. On the recording sheet, *Air Takes up Space,* have the students draw in the boat and the water after each experience.

10. Fill the plastic bag with air and place the mouth of the bag over the mouth of the gallon-size jar.

Secure the bag with a large rubber band around the bag and the neck of the jar. Ask the students what is in the bag and what is in the jar.

11. Invite several students to try to push the plastic bag into the jar. Ask them to explain why they couldn't push the bag into the jar.

12. Have students compare what they learned about the upside cup and the plastic bag in the jar.

13. Invite students to record their explanations on the activity sheet *Air Bags.*

Discussion

1. What did you learn when you put the upside-down cup in the water?

2. What did you learn when you tried to push the bag into the jar?

3. What do you know about air? [Air is everywhere, even in the jar that seemed empty.]

4. Why is it a good thing that we have air on Earth? [All living things need air to live.]

5. How can you show your families that there is air if they can't see it?

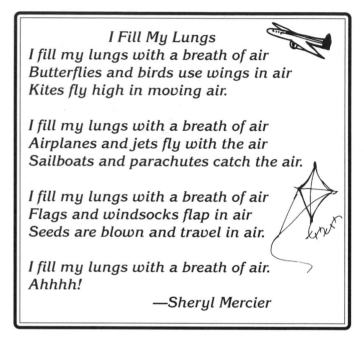

I Fill My Lungs
I fill my lungs with a breath of air
Butterflies and birds use wings in air
Kites fly high in moving air.

I fill my lungs with a breath of air
Airplanes and jets fly with the air
Sailboats and parachutes catch the air.

I fill my lungs with a breath of air
Flags and windsocks flap in air
Seeds are blown and travel in air.

I fill my lungs with a breath of air.
Ahhhh!
—Sheryl Mercier

Extension

Read the poem *I Fill My Lungs.* Use the poem and students' ideas to make a list of how we use air in our daily lives. Have each student choose one idea and draw a picture to illustrate. Arrange the pictures on a chart or bulletin board with the main concepts in the middle.

Air takes up space.
Air has weight.
Air is invisible.
Air is all around.
We use air.

Air Takes up Space

Draw in the boat and water after each step.

Name: _____

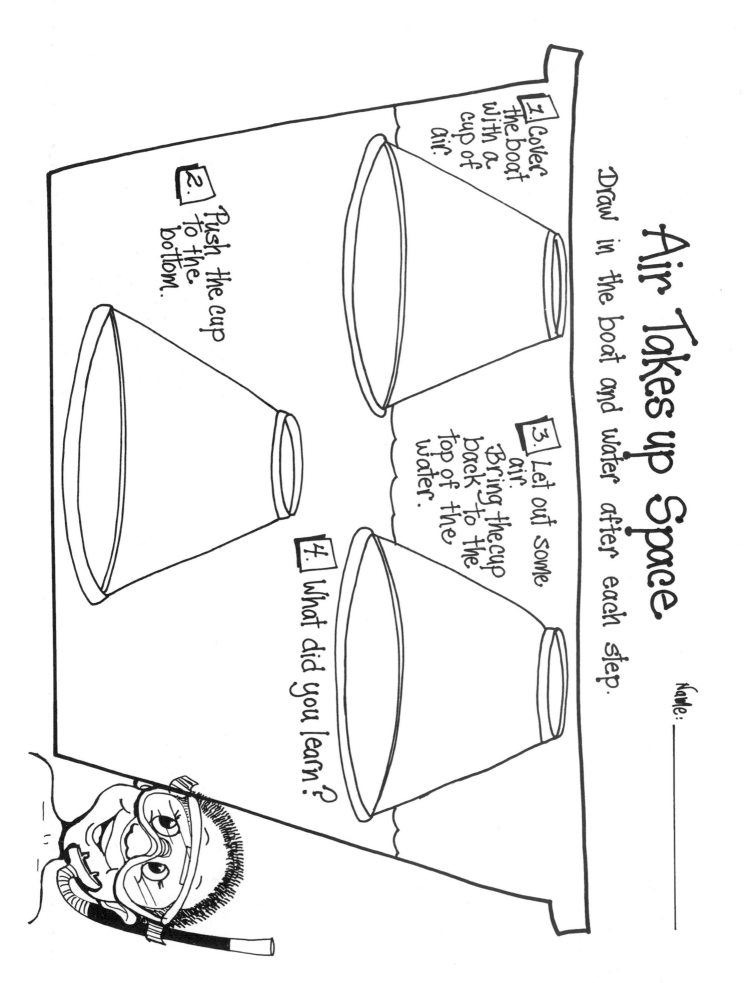

1. Cover the boat with a cup of air.

2. Push the cup to the bottom.

3. Let out some air. Bring the cup back to the top of the water.

4. What did you learn?

Air Bags

Try to push the
bag into the jar.

Draw what happened.

Try to explain what happened.

119

The Wind Blows

Topic
Wind

Key Question
What can we observe about the wind?

Focus
The students will observe and measure how fast the wind is blowing.

Guiding Documents
NCTM Standards
- *Develop the process of measuring and concepts related to units of measurement*
- *Make and use measurements in problems and everyday situations*

NSE Standards
- *Weather changes from day to day and over the seasons. Weather can be described by measurable quantities, such as temperature, wind direction and speed, and precipitation.*

Project 2061 Benchmarks
- *Tools such as thermometers, magnifiers, rulers, or balances often give more information about things than can be obtained by just observing things without their help.*

Math
Measuring
Ordering

Science
Earth science
    wind

Integrated Processes
Observing
Comparing and contrasting
Collecting and recording data
Classifying

Materials
Scissors
Glue
Rulers
Lightweight ribbon or crepe paper streamer (60 cm per student)
Small pieces of paper, styrofoam, bits of material, small pieces of wood and plastic
Measuring tools (see *Management*)

Background Information
Wind is moving air. When the air moves, you can feel it against your face. Wind can move large sailboats across the water. It can turn windmills to pump water. The wind can provide electrical power. Wind can blow so softly that it is not noticed, or it can blow so hard that it can damage buildings, trees, and ships. Wind can cause erosion by blowing away topsoil or by wearing away rock with sharp grains of dust.

The air moves because it has been warmed by the sun. Wind is caused by the uneven heating of the air by the sun. As the sun heats the air, it expands and rises. Air from cooler areas then rushes in to replace the heated air. Cold air is denser and has higher pressure; warm air is less dense and has lower pressure. Air flows from areas of high pressure to areas of low pressure.

We can't see air or wind, but we can see things that the wind is moving: particles of dirt or sand, leaves on the trees, the clouds scurrying across the sky, sailboats moving across the lakes, and the flag boldly waving. Students will not actually be measuring the speed of the wind (distance/time), instead they will be observing the movement of a ribbon in the wind and establishing their own scale for how fast the wind is blowing.

Management
1. Beforehand, cut tissue paper into strips one-inch wide. These will be taped on the end of a ruler.
2. Urge students not to leave any litter from their investigations outdoors.
3. The tools used for measuring should be developmentally appropriate. For older students, meter sticks or trundle wheels may be suitable. For younger students, colored paper cut and taped in meter lengths with no numbers may work better.

Procedure
Exploring Wind–1
1. On a windy day encourage the students to look out the window, or if they have just come in from outside, ask them, "Do you think it is windy outside today? Why do you think that?" Guide them to explain the objects they see being blown by the wind. Encourage them to explain what their opinion of a "windy" day is.
2. Discuss with the students what they know about wind. Can they see it?... hear it?... taste it?... feel it?... smell it?
3. Take the students outdoors. Give them a tissue and have them release it and run to keep up with it.

4. Have pieces of paper, small pieces of a foam cup, a feather, a leaf, and small pieces of wood or plastic and ask the students to guess which ones will be moved the farthest by the wind. Allow time for the students to order the objects according to the distance they think they will be moved by the wind. On a windy day have them test the objects. Ask them to determine how they could make "fair tests" and establish some class rules. An example might be to hold the object over their head, let it go, mark by sticking a pencil in the ground where it lands, and measure the distance.

5. Urge the students to perform three trials. Ask them to explain how they decided which object was carried the farthest. Record on the sheet *The Wind Blows*.

6. Take the students outdoors and have them run in different directions. Ask them how the wind feels against their bodies. Ask them if they can move more easily with the wind or against the wind. Encourage them to explain their answer.

7. Have them describe the sounds they hear when the wind blows.

Wind Speed

8. Invite the students to brainstorm words to describe the action of the wind: gentle, light, breezy, mild, moderate, blustery, whipping, strong, harsh, and dangerous.

9. Ask the students if they can figure out how fast the wind is blowing? [look at the school flag, watch the trees and leaves blowing, or a loose piece of paper being blown] Help the students make their idea of a wind scale. For example:

   No wind—the flag hangs limply, the trees are still

   Light wind—the wind felt on face, the leaves are barely moving

   Moderate wind—light flag extended, small branches move, pieces of paper and trash are blown around

   Strong wind—large branches move, singing heard in wires

   Dangerous wind—difficult to walk against wind, small branches broken off trees

10. Have the students make their own wind measurer by taping a 60 cm length of ribbon or crepe paper to the end of a ruler.

11. While indoors, have the whole class stand in a circle holding their wind measurers up above their heads. Guide everyone to observe that if they stand still, the ribbon or crepe paper streamer hangs down.

12. Tell the students to walk in a circle so they create their own air movement or wind. Guide the students to observe what happens to the wind measurers when they move faster or more slowly. Ask them what they would have to do to make their streamers stay still or stand straight out.

13. Take the students outdoors during different wind conditions and have them observe the action of the wind measurers as they stand still.

14. Discuss how certain types of winds can be helpful or harmful to us. Be sure not to arouse fears of tornadoes, hurricanes, and other extreme conditions.

15. Have the students make a pinwheel, see *Make a Pinwheel* sheet.

16. Let the students test their pinwheel indoors by blowing on the wheel.

17. Take the pinwheels outdoors and let the students test the spin. Have them see if the wind is as strong as their lungs. Urge them to see if they can use the pinwheel to determine wind direction. Ask them if they think they can make some guidelines to know how fast the wind is blowing by using their pinwheel. Allow time for them to test their ideas.

Discussion

1. How do you know when there is wind?
2. What comes to your mind when you hear, "Today is a windy day?"
3. What do you feel when the wind is blowing on your face?
4. Do you always have air around you? How do you know?
5. Describe how the trees move in the wind.
6. Is it correct to say that we see the wind blowing? What do we see? [things blown by the wind]
7. Were you able to keep up with your tissue? Did it blow in the air or along the ground?
8. What things did you find on the school ground that were blown by the wind?
9. Describe the action of your wind measurer.
10. What other ways can you tell how fast the wind is blowing?

121

Extensions

1. Display a wide variety of pictures that show wind, or moving air. Have the students take two of the pictures and compare them as to how hard they think the wind is blowing. You may also wish to have the students decide which pictures show useful wind (as perhaps sailboat sailing), and which shows a harmful wind (a destructive storm).
2. Use the activity *Huff and Puff,* from the AIMS publication *Spring into Math and Science.*
3. Take the students for a walk on a windy day. Encourage them to look for signs of the wind. Ask questions such as these: Are the trees moving in the wind? How do they move? Do the leaves move? What else on the tree is moving? Does the wind in the trees make a sound? What do you think is making that sound?

Curriculum Correlation

Language Arts
1. Read *Gilberto and the Wind* by Marie Hall Ets. Viking Press. NY. 1963. Discuss with the class how much fun Gilberto had playing in the wind. Suggest that the students might want to make their own book describing their own adventures with the wind.
2. Read the Winnie the Pooh adventure, *The Blustery Day* by A.A. Milne.
3. On a windy day, ask the students to observe what the wind is blowing: grasses, leaves, paper, students' clothes. Have the students suggest words that describe the movements. Encourage the students to move to show how they would explain the words.

Art
Drop paint onto slick paper. Blow air through a straw to make wild, windy paint pictures. Blow the paint so that it moves in different directions to create a painting. The paper can be used to make gift wrap.

Home Link
Have students take home their wind measurers and explain how to use them to their parents. Suggest that parents and students keep a record for a week of how fast the wind blows each day.

The Wind Blows

You Need:

leaf

feather

piece of paper

piece of foam cup

Do This:

1. Hold the object.
2. Let go.
3. Mark and measure.

Which one will the wind carry farthest?

Object	Measure			Order
	Drop 1	Drop 2	Drop 3	
1.				
2.				
3.				
4.				

123

Make a Pinwheel

You will need: pencil with eraser, push pin, colors, colored paper, tape, glue

Do this:

1. Color the pinwheel pattern.

2. Cut it out and paste on colored paper.

3. Cut on the inside lines.

4. Bring each corner to the center. Tape.

5. Push the pin through the midde of the pinwheel into the eraser.

6. Test the wind outside with your pinwheel.

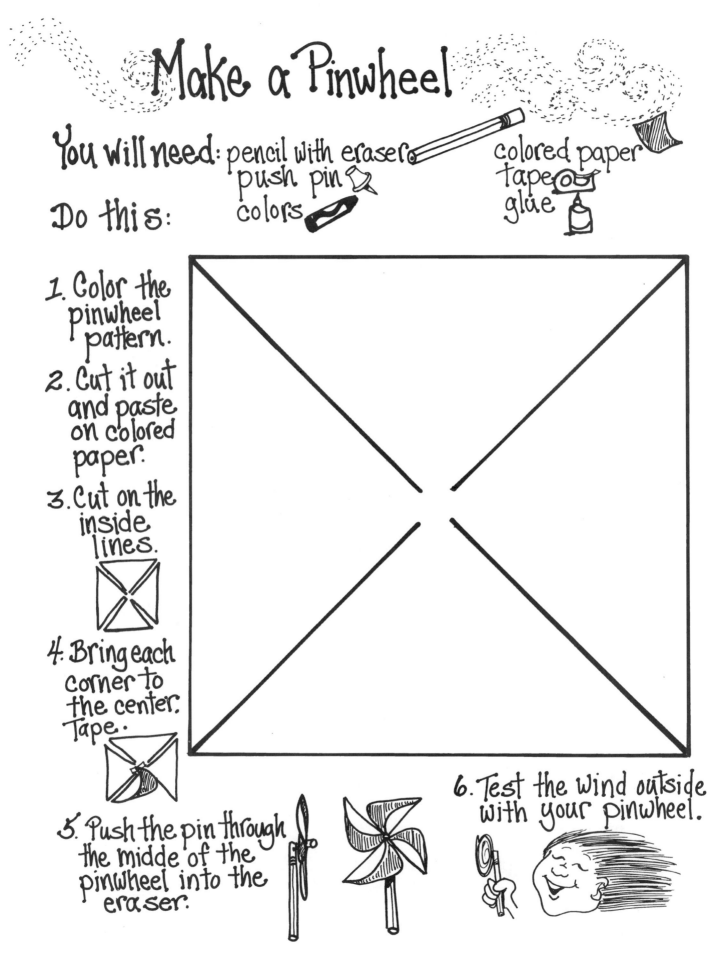

Which Way?

Topic
Wind direction

Key Question
How can you tell from which direction the wind is blowing?

Focus
Wind vanes are tools used to measure wind direction.

Guiding Documents
NSE Standards
- *Weather changes from day to day and over the seasons. Weather can be described by measurable quantities, such as temperature, wind direction and speed, and precipitation.*

Project 2061 Benchmarks
- *People can often learn about things around them by just observing those things carefully, but sometimes they can learn more by doing something to the things and noting what happens.*
- *Tools are used to do things better or more easily and to do some things that could not otherwise be done at all. In technology, tools are used to observe, measure, and make things.*
- *A model of something is different from the real thing but can be used to learn something about the real thing.*
- *Some events in nature have a repeating pattern. The weather changes some from day to day, but things such as temperature and rain (or snow) tend to be high, low, or medium in the same months every year.*

Science
Earth science
   wind direction

Integrated Processes
Observing
Classifying
Predicting
Applying

Materials
Scissors
Glue
Drinking straws
Straight pins
Paper clips
Pencils with erasers
Tagboard
Tape
3" x 5" note cards

Background Information
Wind is moving air. We usually think of it as air that moves along Earth's surfaces. As the sun heats the air, it expands and rises. Cooler, denser air moves in to take its place. This movement of air is wind.

Wind direction is the direction from which the wind comes. We can usually tell from which direction the wind is blowing by watching things as they move. A wind vane is used to indicate the direction from which the wind is blowing. It does this by pointing into the wind.

A wind vane consists basically of an arrow on a pivot. The wind blows the larger end of the arrow away from the source of the wind, causing the smaller, pointed end to point in the direction from which the wind is blowing. For many years people have referred to wind vanes as weather vanes probably because they know the readings are closely related to weather changes.

Winds are named for the direction from which they come. A north wind is blowing from the north to the south. A wind vane points into the wind, so if it points to the north, the wind is a north wind.

Procedure
1. On a windy day, instruct the students to look out the window and to name as many things as they can that tells them that the wind is blowing. [leaves moving, flag flying, movement of the low clouds, dust and dirt blowing, papers and trash moving on the ground] Ask them if they can point in the direction from which the wind is coming.

2. Take the students outside and ask them to determine from which direction the wind is blowing. If they need assistance, suggest that they wet a finger and hold it up. Ask them if one side of their finger feels cooler than the other. Inform them that this is the direction from which the wind is blowing. If they have their wind measurers from the activity *The Wind Blows*, invite them to hold them up. Allow the students time to determine that the wind is blowing from the direction opposite the tail of the ribbon. Ask the students to turn so they face the wind.

3. Play the game *Simon Says* using wind directions. For example, Simon says, turn so the wind blows in your face. Turn so the wind blows on your back. Simon says, turn so the wind hits your right side, etc.
4. Explain to the students that the direction of the wind is shown by a wind vane. Describe a wind vane and ask if anyone has seen one. [on top of barns, houses, and at a weather station]
5. Take the students inside. Inform them that they are going to follow the directions on the sheet *Make a Wind Vane* so they can make their own wind vane.
6. After the students have finished constructing their wind vanes, test them with an electric fan.
7. Take the students and their wind vanes outside to see how they work.
8. After they have seen a wind vane work, ask them if they can think of people who need to know the direction of the wind. [pilots, sailors, parachutists, hunters, air traffic controllers, farmers, weather forecasters]

Using Air

9. Brainstorm with the students all the things that fly in the air or use wind and air currents to move. [birds, insects, kites, balloons, clouds, seeds, gliders, etc.] Explain to the students that we use the wind in many ways: to turn windmills and pinwheels, to dry clothes, to fly kites, etc.
10. Record all the ideas and group them into natural or human-made categories. Have the students bring pictures from home of these things or draw pictures and make a *Using the Wind* bulletin board.

Discussion

1. Why is it important to know what direction the wind is blowing?
2. As you observed the wind, could you tell without using a wind vane what direction the wind was blowing?
3. Explain how a wind vane works?
4. Where do you think wind vanes would be used? Why would they be important at an airport?
5. What things do you like to do on windy days? What things do you like to do on calm days? Are any of them the same as what you do on windy days?

Extension

Another type of instrument used to indicate wind direction is a wind sock. These are often used at small airports. The wind sock fills with air and indicates the direction of the wind and the speed of the wind. The more horizontal the stocking, the stronger the wind is.

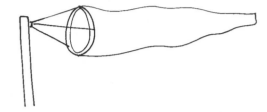

A wind sock is like a sock with a hole cut in the toe, the wind blows through the sock. There are many colorful wind socks on the market today. Hang one outside the classroom so that students can observe the speed of the wind as measured by the wind sock.

Curriculum Correlation

Language Arts

1. Show a picture of people in a very strong wind. Have the students write a story about the picture.
2. Have the students write a creative story about wind. An example of a writing prompt: *One day, I was sitting all by myself when the wind came along and lifted me high over the trees. I felt lighter than air...* Finish the story and tell what you saw, how you felt, and how you got back.

Literature
Bauer, Caroline Feller. *Windy Day.* J. B. Lippincott. NY. 1988.

Home Link

Tell the students to take their wind vanes home and teach their families about wind direction.

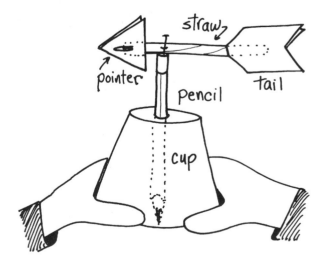

Make a Wind Vane

You will need:
pencil with eraser
cup
pin
straw
tag board
paper clip

Do This:

1. Trace and cut out the pattern on tagboard or stiff paper.

   Make 2 tails and 2 pointers.

2. Put a paper clip on the end of a straw for weight.

3. Push a pin through the straw into a pencil eraser.

   ⅓ length   ⅔ length

4. Glue the 2 pointers together on the paper clip end of the straw.

5. Glue the 2 tails together on the other end of the straw.

6. Push the pencil through the bottom of the cup and use as a holder.

7. Test to find wind direction.

pointer   straw   tail   Pencil   cup

Cloudy Weather

Topic
Clouds

Key Question
What can you observe about clouds?

Focus
The students will observe and classify clouds.

Guiding Documents
NSE Standards
- *Weather changes from day to day and over the seasons. Weather can be described by measurable quantities, such as temperature, wind direction and speed, and precipitation.*
- *The sun, moon, stars, clouds, birds, and airplanes all have properties, locations, and movements that can be observed and described.*

Project 2061 Benchmarks
- *Things move in many different ways, such as straight, zigzag, round and round, back and forth, and fast and slow.*
- *Some events in nature have a repeating pattern. The weather changes some from day to day, but things such as temperature and rain (or snow) tend to be high, low, medium in the same months every year.*
- *The sun warms the land, air, and water.*

Math
Graphing

Science
Earth science
    clouds

Integrated Processes
Observing
Classifying
Predicting
Comparing and contrasting
Communicating
Applying

Materials
Cotton
Blue paper
Glue

Background Information
Clouds are formed from water that has evaporated from rivers, ponds, oceans, and lakes. This water vapor expands and rises. As the water vapor expands, it cools. When the water vapor has cooled enough, it condenses and forms a cloud.

Clouds are masses of small water droplets that float in the air. There is always moisture in the air, but it forms into clouds only when the conditions are right. The tiny droplets of water need the right air temperature and something to cling to, such as dust, salt, or smoke particles. As the water forms into droplets, they seem to float in the sky, moved by the wind.

Each type of cloud carries a message, giving advance notice of coming changes in the weather. Clouds get their shape and form at certain heights because of the movement of air and the amount of water vapor and condensation particles in the air. In 1803, a British scientist used Latin names to identify cloud shape and height above ground: cirrus, stratus, and cumulus. Modern classification has added alto and nimbus. Most clouds can be identified with these names and their combinations.

The names which identify cloud shapes are *stratus* (spread out or layered), *cumulus* (heaped or piled up, lumpy), and *cirrus* (curly, wispy). The prefixes *alto* (middle level) and *cirro* (high level) indicate height above ground. *Nimbus* and the prefix *nimbo* refer to clouds that produce precipitation.

S*tratus* (spread out) are relatively low level, layered clouds that often cover the sky with a blanket of gray clouds. They may be found spread out, anywhere from the ground's surface up to a height of about one mile. On the ground, these clouds are called fog. Stratus clouds often indicate that rain is coming.

Middle-altitude clouds use the prefix *alto*. *Altostratus* clouds are located higher than stratus, but have a similar formation. *Altocumulus* (high heaps) are located at approximately the same height, but have the cumulus shape. They appear as small cloud patches arranged in rounded heaps of whitish masses.

The highest clouds are called *cirrus* (curl of hair). Cirrus clouds are the high, wispy white clouds made up of feathery ice crystals. They indicate a change in the weather. The prefix *cirro* is used to identify high level clouds of different shapes, such as *cirrocumulus* and *cirrostratus*.

Cumulus (heap or pile) are puffy, white clouds with flat bottoms that indicate fair weather. If they are building, it may turn stormy. *Cumulonimbus* (dark rain clouds) are clouds that can stretch up thousands of feet from their base and contain huge amounts of moisture. They are heaped-up piles of clouds. These kinds of clouds are often called thunderheads. Thunder, lightning, and rain often come from them.

Management

1. This activity should be done over a period of time so that students can observe various cloud formations.
2. Keeping a special calendar with illustrations of the clouds and other weather observations for each day may help students to see that clouds can be used as predictors of weather.
3. The illustration of the major types of clouds (*Cloud Types*) is included for your information. A transparency of the *Cloud Types* page can be made and used, if desired, when students begin to talk about the classification of cloud shapes and height.
4. The focus of this lesson is NOT on learning the names of the different clouds. "Naming" should not be considered an indication of "knowing" in this lesson. Instead, the focus is one of observation. Students should observe that clouds have different shapes ("clumpy, straight, wispy") and appear to be at different heights. If the students desire to know the names, these should be considered incidental.
5. If the names of the clouds are not used, alter the graph to include only the illustrations of the different formations.

Procedure

1. Be certain that students have had the opportunity for focused observation of different types of skies: a clear day, a partly cloudy day, a cloudy day.
2. On a partly cloudy day, take the students out on the school grounds to sit down and watch the clouds. Tell them to use their imaginations and describe what shapes they can see. Urge them to continue their descriptions as the clouds change shape. Ask them why they think the clouds change shape. Ask them to share other observations.
3. Back in the classroom, have each student complete the sentence "A cloud looks like..." The answers can be recorded on the board or on a piece of butcher paper cut in the shape of a cloud.

4. On another cloudy/partly cloudy day, have them follow up by doing the recording sheet *Clouds*.
5. Use the students' illustrations to sort clouds into groups such as *puffy clouds*, *straight layered clouds*, *feathery or wispy clouds*. Tell students that people who study clouds also group them by how high they are.
6. Ask the students what they think clouds are made of. After allowing time for discussion, ask the students if they think they have ever created a cloud. If necessary, suggest that they think about their breath on cold days. Allow them time to realize that they make clouds on cold days by breathing out warm, moist air into the cold outdoor air. As the air cools, it briefly forms a little "cloud." Ask them if they have ever seen a "cloud" at the grocery store when they open the freezer door.
7. Tell the students that they may be able to see a cloud in a bottle if they look closely. Inform them that you will make the cloud because it requires hot water and you don't want them to be burned. Point out that the hot water is like their warm moist breath. Pour about a cup of hot water into an empty, clear two-liter bottle. Shake the bottle vigorously to warm the inside of the bottle. Pour out most of the water, leaving only about one capful. Place the bottle in front of the students and screw the cap securely in place. Have the students look through the bottle at the light in order to see the wisps of vapor as it is produced. Squeeze the bottle, then relax your grip.

A cloud should form. If not, repeat the steps but before putting on the cap, strike a match and throw it into the bottle. (The smoke provides particles for condensation.) Squeeze and then relax. Ask the students what they see happening in the bottle.

8. Use blue poster paper to make a sky on a wall or bulletin board. Assign different cloud types to the students so that all are not the same. Distribute blue construction paper. Have them use cotton balls glued to construction paper to make the different kinds of clouds. (If some assistance is needed, suggest that they stretch out the cotton to show the wispy clouds (cirrus) with blue sky showing through, bunch up the cotton for the puffy clouds (cumulus), stretch the cotton for layered clouds (stratus). Create a sky scene, putting the stratus close to the ground and cirrus up high.

9. Suggest that students become cloud watchers to see what story clouds have to tell. Inform them that weather watchers often use clouds to predict what kind of weather is coming.

10. Have students keep track of the *Cloud Types* by graphing the clouds in the sky each day for a month. Challenge them to try to predict the weather by observing the clouds.

Discussion

1. Are there always clouds in the sky? Are there clouds today?
2. Describe the shape of the clouds. How high are they, high, medium, or low?
3. Have you ever been in a cloud that is on the ground? Describe it. How did it feel to walk through it?
4. Have you ever watched the clouds move? What makes them move?
5. What exciting shapes have you observed in the clouds?
6. When there are clouds in the sky, what happens to the sun? [The sun is still there but is sometimes blocked from our view.]
7. When is it brighter outside, when there are clouds or when there are no clouds? Why do you think this happens?
8. If you were asked to predict the weather for the next day just by looking at the clouds, how would you do it?
9. What are clouds made up of?

Extension

Research folklore that is connected with the appearance of different kinds of clouds, such as "Red clouds in the morning sailors take warning. Red clouds at night, sailors delight."

Curriculum Correlation

Language Arts

1. Read the book, *It Looked Like Spilt Milk* by Charles G. Shaw. Harper Trophy. NY. 1947.
2. Discuss with the students what shapes they have seen in the clouds. Have the students write a cloud book. Each page is illustrated to show a shape in the clouds that the students saw.

Cumulonimbus
wind
thunderstorms

Cloud
Types

Cirrus
ice crystal clouds
change in weather

Cumulus
white, puffy
fair weather

Stratus
low, grey   rain or snow

Fog

Clouds

Name: _____

1. Draw a picture of the clouds you see.

2. Describe the clouds you see.

3. What do you think makes clouds?

132

Cloud in a Bottle

1. Observe the cloud inside the 2liter bottle.
2. Draw what you see.
3. Tell how to make a cloud inside of a bottle in 3 steps.

First, _____
_____
_____

Second, _____
_____

Third, _____
_____
_____

water vapor   dust

Cloud Types

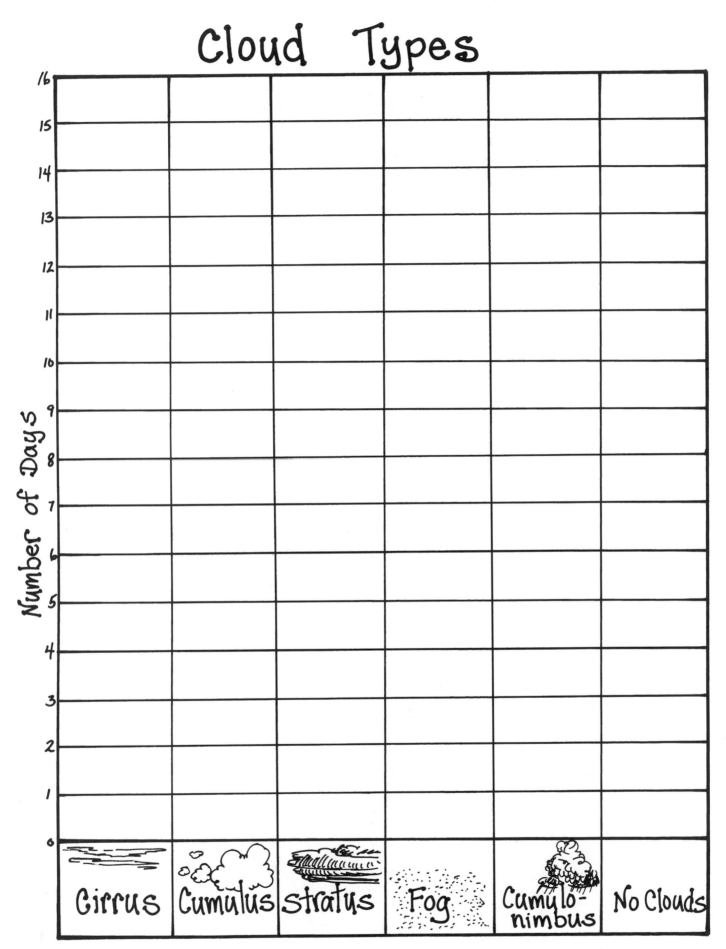

Number of Days

16
15
14
13
12
11
10
9
8
7
6
5
4
3
2
1
0

Cirrus | Cumulus | stratus | Fog | Cumulo-nimbus | No Clouds

Watching the Weather

Topic
Observing the weather

Key Question
What is the weather like today?

Focus
Students will observe and record weather conditions over a long period of time.

Guiding Documents
NCTM Standards
- *Collect, organize and describe data*

NSE Standards
- *Weather changes from day to day and over the seasons. Weather can be described by measurable quantities, such as temperature, wind direction and speed, and precipitation.*
- *The sun, moon, stars, clouds, birds, and airplanes all have properties, locations, and movements that can be observed and described.*

Project 2061 Benchmarks
- *Describing things as accurately as possible is important in science because it enables people to compare their observations with those of others.*
- *Some events in nature have a repeating pattern. The weather changes some from day to day, but things such as temperature and rain (or snow) tend to be high, low, or medium in the same months every year.*
- *Simple graphs can help to tell about observations.*
- *Some things are more likely to happen than others. Some events can be predicted well and some cannot. Sometimes people aren't sure what will happen because they don't know everything that might be having an effect.*

Math
Counting
Graphing

Science
Earth science
    meteorology

Integrated Processes
Observing
Comparing and contrasting
Predicting
Inferring
Collecting and recording data

Materials
Butcher paper
Crayons, markers, or colored pencils
Glue

Background Information
Weather affects our lives in many ways. The type of clothing we wear depends largely on the weather. Many businesses are affected by weather, good and bad. Farmers need rain to water crops and sunshine to make them grow. They need clear weather to plant and harvest crops. Rain, snow, fog, and wind can disrupt transportation. Inclement weather even affects children's plans for outdoor recreation.

Weather is the condition of the atmosphere at a certain time in regards to the amount of sunlight, precipitation, clouds, and range of temperature. The weather can be windy, rainy, cold, hot, sunny, or a combination of things.

Weather can be very different all across our nation. The ocean, mountains, and wind patterns create various weather systems. Young children tend to think that the weather they experience today is the same everywhere. It is a challenge to help them realize that weather varies greatly from region to region.

Management
1. The large weather symbols could be printed on different colors of paper, then cut apart and kept in small boxes or plastic bags.
2. Make a large weather calendar chart on poster board or with yarn on a bulletin board.

Procedure
1. Begin the discussion by asking the students to describe the day's weather. Go outside if necessary. Have them describe what the weather was like yesterday. Ask them how the weather affects them.
2. Use the recording sheet *Observe the Weather*. Tell the students to use their senses to observe the weather.
3. Discuss how they or their parents decided what the weather would be like today. For example: Did they look out the window to see if the sun were shining, raining, snowing, etc. Or did they listen to the weather forecast?
4. Brainstorm words to describe all sorts of weather. (foggy, rainy, snowy, cloudy) List these on the board or on a large cloud-shaped piece of butcher paper.
5. Explain that weather forecasters keep records of daily weather to help them predict the weather for the following days. Tell the students that for the next

month, the class will use some weather symbols to make a record of the weather on a class calendar chart. Show and explain each of the symbols.

6. At the same time each day, ask the students which symbol should be posted to best represent the day's weather. Post the day's symbol on the class calendar chart.

7. After a few days of observing, ask the students to predict what they think the weather will be like tomorrow. Have the class vote and post the most frequent guesses next to the chart. Compare their predictions with the actual weather conditions on the following day.

8. On the student sheet *Watching the Weather Graph*, record the number of days that each type of weather occurred. Keep records for more than one month (or season) and compare the number of clear days, rainy days, foggy days, etc.

9. Have students record the week's weather from the class calendar onto their chart *Watching the Weather*. They can draw or write their observations on the chart.

10. Use the recording sheet *My Favorite Weather* as a culmination for weather observation.

Discussion

1. What is a word you would use to describe the weather today?

2. How many days did we keep a record of the weather?

3. How many clear days did we have during the month? How many days during the month were cloudy? Did we have more clear days than cloudy? How do you know?

4. Looking at the graph, what kind of weather did we have the most of?

5. Did we have any rain (or snow) during the month? If so, how many days?

6. What was the weather like today? What was it like a week ago? Why is it easier to tell what it was like a week ago than to predict what it will be next week?

7. What type of clothing would you wear for a hot, sunny day? What would you wear on a cold, wet day? How does clothing color, weight, and thickness differ with the kind of weather we are having?

8. What kind of weather do you like the best? Why? What do you do on those days?

Extensions

1. Weather affects all life on Earth. Discuss the danger of electrical storms and procedures to stay safe during dangerous weather. Make a class plan for any severe weather in your area.

2. Make some paper dolls, then make clothing to match the local weather. Tell the students they are to pick the clothes for the dolls, dress them according to the day's weather and post them on the board near the weather calendar for the day.

Curriculum Correlation

Art
The students can create pictures to match the weather outside by painting bright sunny days with bold tempera colors, rainy days with water color washes.

Math
Use the results of the *Weather Graph*. Have students make up questions or statements about the graph such as: How many more sunny days were there than cloudy days? How many rainy days and windy days did we have during this month?

Home Links

1. Send home a note asking parents to sit with their child and listen to the weather report (TV, radio, or newspaper). Discuss the weather predictions the following morning in class.

2. Have the students make a home plan with their families for severe weather.

Observe the Weather

Name:_____

Use your senses to observe weather. Draw and write descriptions.

See

Hear

Smell

Feel

Watching the Weather

Name _____

	Monday	Tuesday	Wednesday	Thursday	Friday
Week 1					
Week 2					
Week 3					
Week 4					

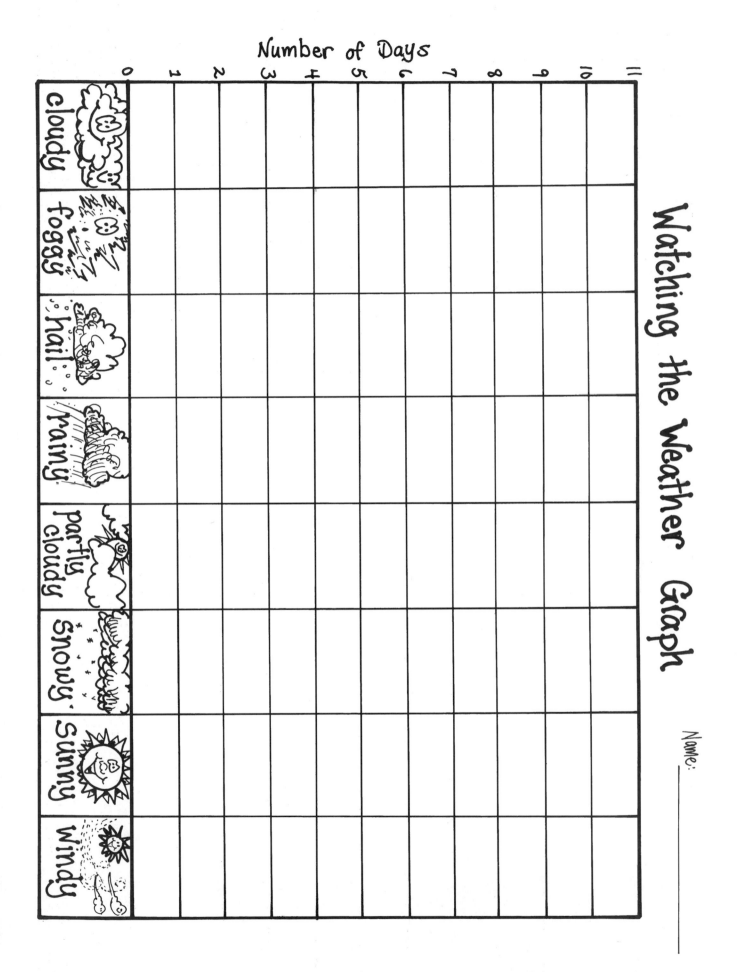

Number of Days

Watching the Weather Graph

Name: _____

cloudy
foggy
hail
rainy
partly cloudy
snowy
sunny
windy

0 1 2 3 4 5 6 7 8 9 10 11

My Favorite Weather

1. Draw a picture of your favorite type of weather. Put yourself in the picture.

2. Write words to describe your favorite weather.

Air Temperature

Topic
Using a thermometer

Key Question
How does air temperature affect a thermometer?

Focus
Students use a thermometer to measure air temperature.

Guiding Documents
NCTM Standards
- *Make and use estimates of measurement*
- *Develop the process of measuring and concepts related to units of measurement*

NSE Standards
- *Weather changes from day to day and over the seasons. Weather can be described by measurable quantities, such as temperature, wind direction and speed, and precipitation.*
- *Simple instruments, such as magnifiers, thermometers, and rulers, provide more information than scientists obtain using only their senses.*

Project 2061 Benchmarks
- *Tools like thermometers, magnifiers, rulers, or balances often give more information about things than can be obtained by just observing things without their help.*
- *Some events in nature have a repeating pattern. The weather changes some from day to day, but things such as temperature and rain (or snow) tend to be high, low, or medium in the same months every year.*

Math
Measuring

Science
Earth science
  temperature

Integrated Processes
Observing
Predicting
Comparing and contrasting
Collecting and recording data
Communicating

Materials
Immersion thermometer
Small buckets
Masking tape, optional
Permanent colored pens, optional

Background Information
Temperature tells us how hot or cold something is. Most people are familiar with the idea of temperature. The temperature of our bodies is important to our health; the temperature of the air is given in the weather report. Our experience tells us that the warmer something feels, the higher the temperature is likely to be.

A thermometer is an instrument whose size, shape, or some other feature changes when its temperature changes so that it can be used to measure temperature. The most common thermometers are those that have an expanding column of mercury or colored alcohol. (In this activity, only alcohol thermometers should be used.)

Measuring temperature can be exciting for students. Just to see the red liquid move up and down seems like magic to them. In this activity, it may be enough that young students observe the relative movement of the colored alcohol as it rises to indicate something is warmer and falls when something is cooler. For older students, attention to the scaling may be appropriate.

Management
1. **CAUTION: Always use alcohol-filled thermometers with young students.**
2. *What's Hot and What's Not* (*AIMS*, Volume 9, Number 9) should be done prior to this activity so that students realize the relative position of the alcohol in the thermometer as it indicates hot or cold. Relative position may be as far as this lesson needs to go for younger students.
3. If students are only looking at relative temperature, cover the thermometer's scales with masking tape. Use permanent pens to color in areas of hot (red), medium (yellow), and cold (blue).
4. If older students have not learned how to read a thermometer, you may wish to use the lesson *What is the Temperature?* in the AIMS publication *Primarily Physics* before you proceed with this lesson.

Procedure
Learning to use the thermometer
1. Instruct the students to look out the window and see if they can tell if it is hot or cold outside. Ask them what they would be looking for to tell them what the temperature is.
2. Ask students to describe what the words *hot* and *cold* mean when referring to air temperatures.
3. Ask them if they have seen a thermometer and if they know how it is used to measure temperatures.
4. Give each group a thermometer. Have the students examine the thermometer and discuss their observations.

5. Explain to the students that the thermometer has colored alcohol in the bulb that expands (takes up more room) when it gets warm or hot and the liquid goes up the tube. Encourage them to offer ways they can get the alcohol to go up the tube.

6. Ask them what they think will happen if they put their thumb on the bulb of the thermometer. Allow time for them to test their predictions. Ask them what they know about the temperature of their finger. [If the alcohol rose in the thermometer, their thumb was warmer than the surrounding air.] Have them remove their thumbs from the thermometers' bulbs and after a short time ask them what has happened to the column of liquid and why it happened.

7. Provide the students with two buckets of water, one warm and one cool. Tell them to stick one hand in each bucket and note the difference. Have them predict what will happen when they place their thermometers in the two different buckets of water.

8. Direct them to place the bulb ends of their thermometers in the buckets of water. Urge them to observe what happens to the red liquid.

9. When the red liquid stops rising, ask the students which colored band (see *Management 3*) or number the red liquid is next to.

10. Encourage them to compare their results with those of other groups and to record their observations.

Using the thermometer for air temperature
11. Have the students take their thermometers outside and record (student sheet *Measure Air Temperature*) the temperature of the air in the sun and in the shade.

12. Ask the students to explain the differences in the temperatures.

13. Encourage them to find other areas which would have big differences in temperatures, for example, the blacktop and a grassy area. Have the students explain why they think one area will be hotter than the other.

Discussion

1. What do you like to eat or drink that is hot? What is your favorite cold food or drink? How does each of these foods become hot or cold?

2. Do you have thermometers at home? What are they used for? [air conditioning, heater, or furnace]

3. When something is warm, what happens to the thermometer? When something is cold what happens to it?

4. When you were using your thermometer, what did you notice about the red liquid in the tube. [It goes up when it is hot and it goes down when it is cold]

5. When you put your thumb on the bulb of the thermometer, what happened to the red liquid? What temperature were you measuring? [my thumb's]

6. When you put the thermometer in the buckets of water, what temperature were you measuring? [the water's]

7. For thermometers with colored masking tape: Beside which color did the red liquid stop when you put the thermometer into the cup of hot water?

8. For thermometers with colored masking tape: On a cool day, what color of tape would you expect the alcohol column to stop at?

9. When you measured the temperature of the air outside, where did you find the coolest temperature?... the warmest?

10. How can you tell if something is hot or cold without actually touching it yourself?

Extension

Challenge the students to find the hottest and the coldest spot on the school ground. Let them explain why they chose certain areas to investigate.

Curriculum Correlation

Literature
   Maestro, Betsy and Giulio. *Temperature and You.* Lodestar Books. NY. 1990.

Music
   Sing *Just Right* from *AIMS*, Volume 9, Number 9.

Measure Air Temperature

Name: _____

Color in the red to match your thermometer:

Where: _____   Where: _____   Where: _____

What did you find out ?

Bibliography

Mountains

Ansel, Sheri. *Mountains*. Steck-Vaughn. Austin, TX. 1993.

Bailey, Donna. *Mountains*. Steck-Vaughn. Austin, TX. 1990.

Bain, Iain. *Mountains and Earth Movements*. Bookwright Press. NY. 1984.

Bender, Lionel. *Mountains*. Franklin Watts. London. 1988.

Berger, Melvin. *As Old as the Hills*. Franklin Watts. London. 1989.

Bradley, Catherine. *Life in the Mountains,* Scholastic, Inc. NY. 1991.

Bramwell, Martyn. *Mountains*. Franklin Watts. NY. 1994

Hiscock, Bruce. *The Big Rock*. Atheneum Books. NY. 1988.

Lye, Keith. *Mountains*. Silver Burdett. Morristown, NJ. 1987.

Mariner, Tom. *Mountains*. Marshall Cavendish. NY. 1990.

Morgan, Patricia. *A Mountain Adventure*. Troll Associates. Mahwah, NJ. l988.

Updegraff, Robert and Imelda. *Mountains and Valleys*. Penguin Books. NY. 1982.

Vaughn, Jenny. *Mountains*. Schoolhouse Press. Lexington, MA. 1986.

Williams, Lawrence. *Mountains*. Marshall Cavendish. NY. 1990.

Rivers

Bailey, Donna. *Rivers*. Steck-Vaughn. Austin, TX. 1990.

Cheney, Glenn Alan. *The Amazon*. Franklin Watts. NY. 1984.

Cherry, Lynne. *A River Ran Wild*. Harcourt Brace and Co. NY. 1984.

Crisman, Ruth. *The Mississippi*. Franklin Watts. NY. 1984.

Fowler, Allan. *All Along the River*. Children's Press. Chicago. 1994.

Halperm, Shari. *My River*. Macmillan Publishing Co. NY. 1992.

Harmer, Trudy J. *The St. Lawrence*. Franklin Watts. NY. 1984.

Holling, Clancy. *Minn of the Mississippi*. Houghton Mifflin Co. Boston. 1979.

Holling, Clancy. *Paddle to the Sea*. Houghton Mifflin Co. Boston. 1969.

Jeunesse & Bour. *The River*. Scholastic, Inc. NY. 1992.

Mariner, Tom. *Rivers*. Marshall Cavendish. NY.1990.

MacEachen, Sally ed., et al. *The Unfolding River*. Quarto Books. London. 1992.

McCauley, Jane R. *Let's Explore a River*. National Geographic Society. Washington D.C. 1988.

Perceful, Aaron W. *The Nile*. Franklin Watts. NY. 1984.

Santrey, Laurence. *Rivers*. Troll Associates. Mahwah, NJ. 1985.

Schmit, Eleanore. *The Water's Journey*. North-South Books. NY. 1990.

Williams, Vera. *Three Days on a River in a Red Canoe*. Mulberry Books. NY. 1981.

Lakes

Bender, Lionel. *Lake*. Franklin Watts. NY. 1989.

Bramwell, Martyn. *Rivers and Lakes*. Franklin Watts. NY. 1994.

Dineen, Jacqueline. *Rivers and Lakes*. Schoolhouse Press. Lexington, MA. 1986.

Mulherin, Jenny. *Rivers and Lakes*. Franklin Watts. NY. 1984.

Rowlan-Entwistle, Theodore. *Rivers and Lakes*. Silver Burdett. Morristown, NJ. 1987.

Plains

Bouchard, David. *If you're not from the prairie...* Antheneum Books for Young Readers. NY. 1994.
Flint, David. *The Prairies and Their People.* Thomson Learning. NY. 1993.

Deserts

Baker. Lucy. *Life in the Deserts.* Scholastic, Inc. NY. 1990.
Baylor, Byrd. *The Desert is Theirs.* Charles Scribner's Sons. NY. 1975.
Bender, Lionel. *Deserts.* Franklin Watts. NY. 1989.
Brandt, Keith. *Deserts.* Troll Associates. Mahwah, NJ. 1985.
George, Jean Craighead. *One Day in the Desert.* Scholastic, Inc. NY. 1983.
Guiderson, Brenda. *Cactus Hotel.* Henry Holt & Co. NY. 1991.
Johnson, Angela. *Toning the Sweep.* Orchard Books. NY. 1993.
Lerner, Carol. *A Desert Year.* Morrow Junior Books. NY. 1991.
Posell, Elsa. *Deserts.* Children's Press. Chicago. 1982.

Valleys

Goetz, Delia. *Valleys.* William Morrow and Co., Inc. NY. 1976.
Updegraff, Imelda and Robert. *Mountains and Valleys.* Penguin Books. NY. 1982.

Oceans

Baker, Lucy. *Life in the Oceans.* Scholastic, Inc. NY. 1990.
Bramwell, Martin. *Oceans.* Franklin Watts. NY. 1984.
Cole, Joanna. *The Magic School Bus on the Ocean Floor.* Scholastic, Inc. NY. 1992.
Costa de Beauregard, Diane. *The Blue Planet: Seas & Oceans.* Young Discovery Library. Ossining, NY. 1989.
Hoff, Mary King. *Our Endangered Planet, Oceans.* Lerner Publishing Co. Minneapolis, MN. 1991.
Jennings, Terry. *Oceans and Seas.* Marshall Cavendish. Freeport, NY. 1989.
Lambert, David. *Seas and Oceans.* Steck-Vaughn. Austin, TX. 1994.
Lionni. Leo. *Swimming.* Alfred A. Knopf, Inc. NY. 1963.
Palmer, Joy. *Oceans.* Steck-Vaughn. Austin, TX. 1992.
Pearce, Q.L. *Tidal Waves and Other Ocean Wonders.* Julian Messner. Englewood Cliffs, NJ. 1989.
Polking, Kirk. *Oceans of the World.* Philomel Books. NY. 1983.
Twist, Clint. *Seas and Oceans.* Crestwood House. NY. 1991.
Whitfield, Philip. *Oceans.* Viking Penguin, Inc. NY. 1991.

Icecaps and Glaciers

Bramwell, Martyn. *Glaciers and Ice Caps.* Franklin Watts. NY. 1986.
George, Michael. *Glaciers.* Creative Education, Mankato, MN. 1991.
Georges. D.V. *Glaciers.* Children's Press. Chicago. 1986.
Hackwell, John W. *Desert of Ice: Life and Work in Antarctica.* Charles Scribner's Sons. NY. 1991.
Nixon, Hershell. *Glaciers, Nature's Frozen Rivers.* Dodd, Mead and Co. NY. 1980.
Simon, Seymour. *Icebergs and Glaciers.* William Morrow and Co. NY. 1987.
Walker, Sally. *Glaciers: Ice on the Move.* Carolrhoda Books, Inc. Minneapolis, MN. 1990.

Rocks, Sand, and Soil

Arem, Joel E. *Discover Rocks and Minerals.* Publications International, Ltd. Lincolnwood, IL. 1991.

Baylor, Byrd. *Everybody Needs a Rock.* Macmillan Publishing Co. NY. 1974.

Bourgeous, Paulette. *The Amazing Dirt Book.* Addison-Wesley. NY. 1990.

Catherall, Ed. *Exploring Soil and Rocks.* Steck-Vaughn. Austin, Texas. 1990.

Gans, Roma. *Rock Collecting.* HarperCollins. NY. 1984.

Hiscock, Bruce. *The Big Rock.* Macmillan Publishing Co. NY. 1988.

McNulty, Faith. *How to Dig a Hole to the Other Side of the World.* Scholastic, Inc. NY. 1979.

Podendorf, Illa. *Rocks and Minerals.* Children's Press. Chicago. 1982.

Snedden, Robert. *The Super Science Books of Rocks and Soils.* Thomson Learning. NY. 1995.

Watts, Lisa and Jenny Tyler. *The Usborne Book of the Earth.* Usborne Publishing Ltd. London. 1986.

Webb, Angela. *Sand.* Franklin Watts. NY. 1987.

Air and Water

Barrett, Judith. *Cloudy With a Chance of Meatballs.* Atheneum Books. NY. 1978.

Branley, Franklyn M. *Air Is All Around You.* HarperCollins. NY. 1986.

Branley, Franklyn M. *Flash, Crash, Rumble, and Roll.* HarperCollins. NY. 1985.

Calhoun, Mary. *Jack and the Whoopee Wind.* William Morrow and Co. NY. 1987.

Craft. Ruth. *The Day of the Rainbow.* Viking Press. NY. 1989.

de Paola, Tomie. *The Cloud Book.* Holiday Books. NY. 1975.

Jennings, Terry. *Weather.* Gloucester Press. NY. 1990.

Peters, Lisa. *The Sun, the Wind and the Rain.* Henry Holt & Co. New York. 1988.

Rogers, Paul. *What Will the Weather Be Like Today?* Greenwillow Books. NY. 1990.

Schulevitz, Uri. *Rain Rain Rivers.* Farrar, Strauss, Giroux. NY. 1969.

Skofield, James. *All Wet! All Wet!* Charlotte Zolotow. NY. 1984.

Tresselt, Alvin. *Hide and Seek Fog.* Lothrop, Lee, & Shepard Books. NY. 1988.

The AIMS Program

AIMS is the acronym for "Activities Integrating Mathematics and Science." Such integration enriches learning and makes it meaningful and holistic. AIMS began as a project of Fresno Pacific University to integrate the study of mathematics and science in grades K-9, but has since expanded to include language arts, social studies, and other disciplines.

AIMS is a continuing program of the non-profit AIMS Education Foundation. It had its inception in a National Science Foundation funded program whose purpose was to explore the effectiveness of integrating mathematics and science. The project directors in cooperation with 80 elementary classroom teachers devoted two years to a thorough field-testing of the results and implications of integration.

The approach met with such positive results that the decision was made to launch a program to create instructional materials incorporating this concept. Despite the fact that thoughtful educators have long recommended an integrative approach, very little appropriate material was available in 1981 when the project began. A series of writing projects have ensued and today the AIMS Education Foundation is committed to continue the creation of new integrated activities on a permanent basis.

The AIMS program is funded through the sale of this developing series of books and proceeds from the Foundation's endowment. All net income from program and products flows into a trust fund administered by the AIMS Education Foundation. Use of these funds is restricted to support of research, development, and publication of new materials. Writers donate all their rights to the Foundation to support its on-going program. No royalties are paid to the writers.

The rationale for integration lies in the fact that science, mathematics, language arts, social studies, etc., are integrally interwoven in the real world from which it follows that they should be similarly treated in the classroom where we are preparing students to live in that world. Teachers who use the AIMS program give enthusiastic endorsement to the effectiveness of this approach.

Science encompasses the art of questioning, investigating, hypothesizing, discovering, and communicating. Mathematics is the language that provides clarity, objectivity, and understanding. The language arts provide us powerful tools of communication. Many of the major contemporary societal issues stem from advancements in science and must be studied in the context of the social sciences. Therefore, it is timely that all of us take seriously a more holistic mode of educating our students. This goal motivates all who are associated with the AIMS Program. We invite you to join us in this effort.

Meaningful integration of knowledge is a major recommendation coming from the nation's professional science and mathematics associations. The American Association for the Advancement of Science in *Science for All Americans* strongly recommends the integration of mathematics, science, and technology. The National Council of Teachers of Mathematics places strong emphasis on applications of mathematics such as are found in science investigations. AIMS is fully aligned with these recommendations.

Extensive field testing of AIMS investigations confirms these beneficial results.

1. Mathematics becomes more meaningful, hence more useful, when it is applied to situations that interest students.
2. The extent to which science is studied and understood is increased, with a significant economy of time, when mathematics and science are integrated.
3. There is improved quality of learning and retention, supporting the thesis that learning which is meaningful and relevant is more effective.
4. Motivation and involvement are increased dramatically as students investigate real-world situations and participate actively in the process.

We invite you to become part of this classroom teacher movement by using an integrated approach to learning and sharing any suggestions you may have. The AIMS Program welcomes you!

AIMS Education Foundation Programs

A Day with AIMS

Intensive one-day workshops are offered to introduce educators to the philosophy and rationale of AIMS. Participants will discuss the methodology of AIMS and the strategies by which AIMS principles may be incorporated into curriculum. Each participant will take part in a variety of hands-on AIMS investigations to gain an understanding of such aspects as the scientific/mathematical content, classroom management, and connections with other curricular areas. *A Day with AIMS* workshops may be offered anywhere in the United States. Necessary supplies and take-home materials are usually included in the enrollment fee.

A Week with AIMS

Throughout the nation, AIMS offers many one-week workshops each year, usually in the summer. Each workshop lasts five days and includes at least 30 hours of AIMS hands-on instruction. Participants are grouped according to the grade level(s) in which they are interested. Instructors are members of the AIMS Instructional Leadership Network. Supplies for the activities and a generous supply of take-home materials are included in the enrollment fee. Sites are selected on the basis of applications submitted by educational organizations. If chosen to host a workshop, the host agency agrees to provide specified facilities and cooperate in the promotion of the workshop. The AIMS Education Foundation supplies workshop materials as well as the travel, housing, and meals for instructors.

AIMS One-Week Perspectives Workshops

Each summer, Fresno Pacific University offers AIMS one-week workshops on its campus in Fresno, California. AIMS Program Directors and highly qualified members of the AIMS National Leadership Network serve as instructors.

The Science Festival and the Festival of Mathematics

Each summer, Fresno Pacific University offers a Science Festival and a Festival of Mathematics. These festivals have gained national recognition as inspiring and challenging experiences, giving unique opportunities to experience hands-on mathematics and science in topical and grade-level groups. Guest faculty includes some of the nation's most highly regarded mathematics and science educators. Supplies and take-home materials are included in the enrollment fee.

The AIMS Instructional Leadership Program

This is an AIMS staff-development program seeking to prepare facilitators for leadership roles in science/math education in their home districts or regions. Upon successful completion of the program, trained facilitators become members of the AIMS Instructional Leadership Network, qualified to conduct AIMS workshops, teach AIMS in-service courses for college credit, and serve as AIMS consultants. Intensive training is provided in mathematics, science, process and thinking skills, workshop management, and other relevant topics.

College Credit and Grants

Those who participate in workshops may often qualify for college credit. If the workshop takes place on the campus of Fresno Pacific University, that institution may grant appropriate credit. If the workshop takes place off-campus, arrangements can sometimes be made for credit to be granted by another college or university. In addition, the applicant's home school district is often willing to grant in-service or professional development credit. Many educators who participate in AIMS workshops are recipients of various types of educational grants, either local or national. Nationally known foundations and funding agencies have long recognized the value of AIMS mathematics and science workshops to educators. The AIMS Education Foundation encourages educators interested in attending or hosting workshops to explore the possibilities suggested above. Although the Foundation strongly supports such interest, it reminds applicants that they have the primary responsibility for fulfilling *current* requirements.

For current information regarding the programs described above, please complete the following:

Information Request

Please send current information on the items checked:

___ *Basic Information Packet* on AIMS materials
___ *Festival of Mathematics*
___ *Science Festival*
___ AIMS Instructional Leadership Program

___ *AIMS One-Week Perspectives* workshops
___ *A Week with AIMS* workshops
___ Hosting information for *A Day with AIMS* workshops
___ Hosting information for *A Week with AIMS* workshops

Name _____ Phone _____

Address _____
        Street   City   State   Zip

We invite you to subscribe to *AIMS*!

Each issue of *AIMS* contains a variety of material useful to educators at all grade levels. Feature articles of lasting value deal with topics such as mathematical or science concepts, curriculum, assessment, the teaching of process skills, and historical background. Several of the latest AIMS math/science investigations are always included, along with their reproducible activity sheets. As needs direct and space allows, various issues contain news of current developments, such as workshop schedules, activities of the AIMS Instructional Leadership Network, and announcements of upcoming publications.

AIMS is published monthly, August through May. Subscriptions are on an annual basis only. A subscription entered at any time will begin with the next issue, but will also include the previous issues of that volume. Readers have preferred this arrangement because articles and activities within an annual volume are often interrelated.

Please note that an *AIMS* subscription automatically includes duplication rights for one school site for all issues included in the subscription. Many schools build cost-effective library resources with their subscriptions.

YES! I am interested in subscribing to *AIMS*.

Name _____   Home Phone _____

Address _____   City, State, Zip _____

Please send the following volumes (subject to availability):

_____	Volume V	(1990-91)	$30.00	_____ Volume X	(1995-96)	$30.00
_____	Volume VI	(1991-92)	$30.00	_____ Volume XI	(1996-97)	$30.00
_____	Volume VII	(1992-93)	$30.00	_____ Volume XII	(1997-98)	$30.00
_____	Volume IX	(1994-95)	$30.00	_____ Volume XIII	(1998-99)	$30.00

_____ **Limited offer: Volumes XIII & XIV (1998-2000) $55.00**
(Note: Prices may change without notice)

Check your method of payment:

❏ Check enclosed in the amount of $ _____

❏ Purchase order attached (Please include the P.O.#, the authorizing signature, and position of the authorizing person.)

❏ Credit Card   ❏ Visa   ❏ MasterCard   Amount $ _____

Card # _____   Expiration Date _____

Signature _____   Today's Date _____

Make checks payable to **AIMS Education Foundation.**
Mail to *AIMS* magazine, P.O. Box 8120, Fresno, CA 93747-8120.
Phone (209) 255-4094 or (888) 733-2467 FAX (209) 255-6396
AIMS Homepage: http://www.AIMSedu.org/

AIMS Program Publications

GRADES K-4 SERIES

Bats Incredible
Brinca de Alegria Hacia la Primavera con las Matemáticas y Ciencias
Cáete de Gusto Hacia el Otoño con la Matemática y Ciencias
Cycles of Knowing and Growing
Fall Into Math and Science
Field Detectives
Glide Into Winter With Math and Science
Hardhatting in a Geo-World (Revised Edition, 1996)
Jaw Breakers and Heart Thumpers (Revised Edition, 1995)
Los Cincos Sentidos
Overhead and Underfoot (Revised Edition, 1994)
Patine al Invierno con Matemáticas y Ciencias
Popping With Power (Revised Edition, 1996)
Primariamente Física (Revised Edition, 1994)
Primarily Earth
Primariamente Plantas
Primarily Physics (Revised Edition, 1994)
Primarily Plants
Sense-able Science
Spring Into Math and Science
Under Construction

GRADES K-6 SERIES

Budding Botanist
Critters
El Botanista Principiante
Mostly Magnets
Ositos Nada Más
Primarily Bears
Principalmente Imanes
Water Precious Water

GRADES 5-9 SERIES

Actions with Fractions
Brick Layers
Conexiones Eléctricas
Down to Earth
Electrical Connections
Finding Your Bearings (Revised Edition, 1996)
Floaters and Sinkers (Revised Edition, 1995)
From Head to Toe
Fun With Foods
Historical Connections in Mathematics, Volume I
Historical Connections in Mathematics, Volume II
Historical Connections in Mathematics, Volume III
Machine Shop
Magnificent Microworld Adventures
Math + Science, A Solution
Off the Wall Science: A Poster Series Revisited
Our Wonderful World
Out of This World (Revised Edition, 1994)
Pieces and Patterns, A Patchwork in Math and Science
Piezas y Diseños, un Mosaic de Matemáticas y Ciencias
Soap Films and Bubbles
Spatial Visualization
The Sky's the Limit (Revised Edition, 1994)
The Amazing Circle, Volume 1
Through the Eyes of the Explorers:
    Minds-on Math & Mapping
What's Next, Volume 1
What's Next, Volume 2
What's Next, Volume 3

For further information write to:
    AIMS Education Foundation • P.O. Box 8120 • Fresno, California 93747-8120

AIMS Duplication Rights Program

AIMS has received many requests from school districts for the purchase of unlimited duplication rights to AIMS materials. In response, the AIMS Education Foundation has formulated the program outlined below. There is a built-in flexibility which, we trust, will provide for those who use AIMS materials extensively to purchase such rights for either individual activities or entire books.

It is the goal of the AIMS Education Foundation to make its materials and programs available at reasonable cost. All income from the sale of publications and duplication rights is used to support AIMS programs; hence, strict adherence to regulations governing duplication is essential. Duplication of AIMS materials beyond limits set by copyright laws and those specified below is strictly forbidden.

Limited Duplication Rights

Any purchaser of an AIMS book may make up to *200 copies* of any activity in that book for use at *one school site*. Beyond that, rights must be purchased according to the appropriate category.

Unlimited Duplication Rights for Single Activities

An individual or school may purchase the right to make an unlimited number of copies of a single activity. The royalty is $5.00 per activity per school site.

Examples: 3 activities x 1 site x $5.00 = $15.00
9 activities x 3 sites x $5.00 = $135.00

Unlimited Duplication Rights for Entire Books

A school or district may purchase the right to make an unlimited number of copies of a single, *specified* book. The royalty is $20.00 per book per school site. This is in addition to the cost of the book.

Examples: 5 books x 1 site   x $20.00 = $100.00
12 books x 10 sites x $20.00 = $2400.00

Magazine/Newsletter Duplication Rights

Members of the AIMS Education Foundation who purchase the *AIMS* magazine/*Newsletter* are hereby granted permission to make up to 200 copies of any portion of it, provided these copies will be used for educational purposes.

Workshop Instructors' Duplication Rights

Workshop instructors may distribute to registered workshop participants a maximum of 100 copies of any article and/or 100 copies of no more than eight activities, provided these six conditions are met:

1. Since all AIMS activities are based upon the *AIMS Model of Mathematics* and the *AIMS Model of Learning*, leaders must include in their presentations an explanation of these two models.
2. Workshop instructors must relate the AIMS activities presented to these basic explanations of the AIMS philosophy of education.
3. The copyright notice must appear on all materials distributed.
4. Instructors must provide information enabling participants to apply for membership in the AIMS Education Foundation or order books from the Foundation.
5. Instructors must inform participants of their limited duplication rights as outlined below.
6. Only student pages may be duplicated.

Written permission must be obtained for duplication beyond the limits listed above. Additional royalty payments may be required.

Workshop Participants' Rights

Those enrolled in workshops in which AIMS student activity sheets are distributed may duplicate a maximum of 35 copies or enough to use the lessons one time with one class, whichever is less. Beyond that, rights must be purchased according to the appropriate category.

Application for Duplication Rights

The purchasing agency or individual must clearly specify the following:
1. Name, address, and telephone number
2. Titles of the books for Unlimited Duplication Rights contracts
3. Titles of activities for Unlimited Duplication Rights contracts
4. Names and addresses of school sites for which duplication rights are being purchased.

NOTE: Books to be duplicated must be purchased separately and are not included in the contract for Unlimited Duplication Rights.

The requested duplication rights are automatically authorized when proper payment is received, although a *Certificate of Duplication Rights* will be issued when the application is processed.

Address all correspondence to: **Contract Division**
AIMS Education Foundation
P.O. Box 8120
Fresno, CA 93747-8120